FIELD GUIDE to the GEMS and MINERALS of MEXICO

(Exclusive of Baja California)

A guide to the most productive and interesting mineral and gem collecting localities, with a Spanish-English, English-Spanish glossary of gem, mineral and mining terms.

BY
PAUL WILLARD JOHNSON

Price $2.00

SBN 910652-01-5
Published By
GEMBOOKS
MENTONE, CALIFORNIA

CONTENTS

ACKNOWLEDGMENTS

This book would never have been written except for the initial suggestion and encouragement of Charles J. Parsons, C.G., F.G.A.

Fred W. Bailey, Fresnillo Company, Dr. Ivan F. Wilson, chief geologist, and Juan Holguin, assistant superintendent, both of Fresnillo's Naica mine, helped with data and review. Dr. Edwin Roedder, U.S.G.S., contributed data on gypsum inclusions. Dr. Allen M. Bassett, Department of Geology, San Diego State College, critically reviewed the Naica manuscript.

We received invaluable help from George M. Potter and Dr. Ralph L. Miller of the U.S. Embassy, Mexico City. Senior Engineer Raymundo Llamas, Comision de Fomento Minero, Mexico City, kindly wrote introductions to the Comision's district directors. Dr. Zoltan de Cserna, geologist, and Dr. Eduardo Schmitter, mineralogist, National Geological Institute, Mexico City, offered help on collecting localities.

Robert Cuadra, president of the Miner's Association, San Luis Potosi, showed us every courtesy while there. P. J. Cutting, formerly with Pennsalt Chemical Company made possible our visit to the Esperanza fluorite mine.

Engineer Macario Muñoz, Comision de Fomento Minero, Durango, arranged our visit to Cerro de Mercado. Scott J. Williams, Scottsdale, Arizona, provided data and Dr. George Switzer and Paul E. Desautels, both of the U.S. National Museum, read and offered suggestions on the Mapimi section.

For material on Mexican agate, I am indebted to Col. E. M. Barron, Southern Gem Mining Company, and William J. Lyons, Lyko Mineral and Gem, Inc., both of El Paso, Texas.

And, for their abundant help in proofing the manuscript and the pleasure of accompanying them on over 12,000 miles of field trips into Mexico, I thank and gratefully dedicate this effort to my true friends —

Nathalia Crane Black, G.G.
and
Vete George Black, G.G., F.G.A.

PREFACE

As far as we know, this is the first guidebook of Mexico directed specifically to gem and mineral collectors. Literature and information on collecting in Mexico are scant, scattered, and very hard to find. We have tried to gather as much information as possible, condense, and present it in a form that will be useful.

It is obviously impossible for any one person to cover, personally, every area mentioned here. The author has made several trips to Mexico, on which his knowledge of specific areas is based. Other data were obtained from individuals who have visited and collected in Mexico, from letters from collectors in both the United States and Mexico, and from a great mass of published literature, both public and private.

Where specific directions and information on collecting are given, the data are drawn from the author's or someone else's personal experience. Where districts, mines, or areas are described and only the names, mines, location, and formal geologic and mineralogic data are given, the information is generally from published sources.

Arrangement

The book is arranged to be useful for planning trips. The areas described are grouped into five sections. Three sections contain all areas accessible from the three major north-south highways from Nogales, Ciudad Juarez (El Paso), and Piedras Negras (Eagle Pass). The fourth section covers the great east-west volcanic belt across central Mexico; the fifth includes the area south of this belt. See Index Map

No attempt has been made to give general travel information. This can be obtained from many guidebooks. To repeat it here would be to use space better employed to present actual locality information.

Among the best of these guides for the motoring tourist is *Mexico and Central America,* published by the American Automobile Association. It contains all needed information about border formalities, insurance, and general travel. By all means obtain a copy of this book before starting to plan your trip.

Terry's Guide to Mexico, by T. Philip Terry, Houghton Mifflin Company, Boston, Mass., is still the best general descriptive guide to Mexico. The original *Terry's Guide* is out of print and available only at libraries or through used book shops. The editions of 1935 to 1944 are very complete except hotel and similar information which is out of date. A new edition under the same title was brought out in 1962 but it is only a shadow of its former self and no better than several other guides.

John Wilhelm's *Guide to Mexico City* is the best for those wishing specific information about the capital. It is very complete and new editions are brought out frequently.

INTRODUCTION

Before 1902, Dr. George F. Kunz wrote: "Considering the extent of country in Mexico and in the adjoining States of the Central American Republics, and the richness of mineral wealth that must surely exist there, our present knowledge of the occurrence of precious stones is remarkably small." His statement holds true today. The great mineral wealth of Mexico has been established but the only gem minerals that have been exploited in commercial quantities are opal and agate.

The literature on the gemology and mineralogy of Mexico is scanty and scattered. A few districts are well represented, others of equal importance have been entirely neglected. Still more have been covered only from a commercial mining viewpoint. Mexico is the world's largest silver producer. It also produces gold, oil, copper, lead, zinc, antimony, mercury, bismuth, molybdenum, cadmium, tin, tungsten, manganese, arsenic and graphite.

Most of the descriptions of the Mexican

mineral deposits have been written from an economic viewpoint. Consequently, reference is frequent to the common metallic ore minerals such as galena and sphalerite. The rare minerals of interest to collectors are rarely mentioned. It is frequently possible to collect interesting and unusual specimens that are not mentioned in the literature.

Geography and Physiography

Geographically, Mexico lies in both the North Temperate and Torrid Zones. Though actually divided into some sixteen physiographic regions, many of these can be combined into five basic land areas that have controlled not only the climate and physiographic development of the country, but its very history.

There are relatively narrow coastal plains on both east and west. Dominating the country are the great Sierra Madre Occidental and Sierra Madre del Sur which extend along the western side of the country from north to south. The two ranges, for practical purposes, are one mountain system and, in turn, part of the system that extends from Alaska to the southern tip of South America.

A short distance from the east coast, the Sierra Madre Oriental rises abruptly. This great barrier runs from the Rio Grande nearly to the Isthmus of Tehuantepec. Both of these mountain chains are primarily granitic.

Contrary to these two great ranges is a belt of volcanic mountains and ranges that runs from east to west, bisecting Mexico from Cape Corrientes to Vera Cruz. This belt is still active, and volcanoes erupt frequently. The numerous faults that cut the region are the source of most of Mexico's earthquakes. It is in this belt that the majestic volcanic peaks are found.

Between the two great mountain ranges and north of the volcanic belt, is the high central plateau, the heart of Mexico. Nearly all of the great cities except the seaports are on the plateau. The elevation tempers the climate and in spite of most of its being in the Torrid Zone, the weather is nearly always pleasant and mild.

Mexico City is in its own valley in the heart of the volcanic belt at an elevation of 7,349 feet. It is flanked by towering mountains on both sides with the twin snow-capped peaks of Iztaccihuatl (17,454 feet) and Popocatépetl (17,872 feet) dominating the region. The summit of the highway to Toluca is over 11,000 feet and one must climb to over 10,000 feet to go from Mexico City to Puebla.

Rivers rising in the mountains have cut spectacular *barrancas* (canyons) in the plateau, and these and the mountains have dominated travel in Mexico for centuries. Both have presented formidable barriers to the expansion of roads and railroads, and only in recent years has modern engineering constructed a highway system over which tourists and commerce can travel with ease.

The high plateau is composed primarily of Cretaceous formations and detritic and volcanic rocks. The latter are by far the most important mineralogically. Volcanic activity started during the Cretaceous or at the beginning of the Tertiary period. It still goes on today and large areas are covered with recent rocks on the surface.

The high plateau is presumed to be a folded region where a tremendous mass of volcanic rocks came to rest. Andesite lava was first laid down, then valleys were formed by erosion. These in turn gave rise to additional eruptions which produced cumulative layers of rhyolites and dacites.

Planning

Planning of any trip is essential to its success. While service for automotive tourists has greatly expanded in the past few years, and the main highways are well supplied with service stations and other conveniences, you get into primitive country as soon as you leave the main roads.

To start, you will need a tourist card (obtainable from any Mexican consulate) a birth certificate (or other proof of citizenship), and a valid smallpox vaccination certificate. Car permits can be obtained at the border.

Your United States auto insurance may not be valid in Mexico and you definitely should obtain a Mexican policy. Insurance can be obtained at reasonable rates to cover the duration of your stay. La Provincial is a reputable Mexican company. You can also arrange for Mexican insurance through the American Automobile Association or any of its affiliated clubs, at local offices or at points of entry.

As soon as you leave the border, you will be doing business in Mexican currency. The current exchange rate is approximately 12 pesos for one dollar. A peso is worth about eight cents. To convert pesos to dollars, multiply by eight and point off two places, i.e., 100 pesos times eight equals 800. Point off two places from the right and you get $8.00.

In Mexico they use the dollar sign for pesos. When you see the dollar sign ($), it means pesos unless it is specifically qualified with the word *dollars, dolares,* or *dls.* In some tourist centers you may find prices quoted in U.S. dollars but only infrequently.

A Word on Food

Food preparation and handling is different south of the border. Except in the very best hotels and restaurants, beware of ice cream, milk, water, and all fresh fruits that cannot be peeled. Never eat raw fruits and vegetables except those that can be peeled, and such things as melons, or you may wind up with "Montezuma's revenge."

Food in the better hotels in most cities is good. Food in the villages and elsewhere may be dangerous to anyone not accustomed to it. We found it best to take along enough food for the complete trip, as we were far from supermarkets. Good clean markets are to be found in many large cities, however, and most of them carry U.S. brands of canned and packaged goods.

Water also is unreliable. Do not use ordinary tap water for drinking. Bottled water can be obtained almost anywhere and it is pure and safe. Five-gallon bottles of *Agua Electrapura* can be purchased in Mexico City (and some other large cities) for 15 cents plus the cost of the bottle. Mexican beer is among the best in the world. Both beer and bottled soda pop are safe, as they are made from purified water.

It is a good idea to carry some Halazone tablets to use in local water just in case you are caught without a supply of known good water. To be on the safe side, particularly in remote regions, carry a supply of sulfaguanidine and Kaopectate with neomycin in case you have a touch of the *turista.*

Maps and Information

Tourist information and maps can be obtained from offices of the Mexican Government Tourism Department in many United States cities. Or you may write or visit the Department at Paseo de la Reforma 35, Mexico City, D.F.

Information, maps, and folders can be obtained from Pemex Travel Club, Avenida Juarez 89, Mexico City. It has branch offices at 3104 San Bernardo St., Laredo, Texas; Chamber of Commerce Bldg., El Paso, Texas; and 208 Grand Ave., Nogales, Arizona.

Many Tourist Department and Pemex Auto Club maps, brochures, and folders can be obtained from the Mexican consulates.

Two of the best maps of Mexico are the American Automobile Association's road map and Rand McNally's *Imperial Map of Mexico.*

The Goodrich Tire Company (Mexico) publishes *Caminos de Mexico.* The main feature of this book, written in Spanish, is that the principal highways are listed with the kilometer markings from Mexico City. It has many sectional fold-out maps and a master alphabetical listing of cities and towns.

A good combination is to use air charts in conjunction with the best road maps. Air charts are available from U.S. Coast and Geodetic Survey in many major cities or Washington 25, D.C.

A good guide is *Mexico and Central America,* published annually by the American Automobile Association. This directory lists hotels, motels, trailer parks, and restaurants. It also gives general information about the major tourist attractions.

The weekly publication, *This is Mexico,* published in English, is helpful for current events. It is available at most tourist bureaus in Mexico City, etc.

It is advisable to carry a compass in Mexico. If you get lost and have a good map, it can be very handy, as many road intersections are unmarked.

The kilometer markers along the main highways and some of the roads designate the distance from Mexico City.

Maps of various parts of Mexico can be purchased at the American Book Store, Avenida Madero 25, Mexico City 1, D.F. It also has most of the guidebooks to Mexico published in English. English is spoken.

A wide variety of official Mexican government maps can be purchased from Direccion General Meteorologia y Hidrografia, Departemente Mapoteca Avenida Observatorio 182, Tacubaya (a suburb of Mexico City). Available is a folio of rather old state maps that are sometimes useful for finding locations in old references. They are of little value as road maps. The series is available as single sheets of one state or bound in a folio.

The Mexican government also publishes a complete series of maps on a scale of 1:500,000. There are also some sheets of the projected world map series on a scale of 1:1,000,000. Both of these are fairly well printed in multicolor, and are probably the best maps available though road information is not up to date.

A large, multicolored geologic map of Mexico can be obtained from the Geology Department, University of Mexico, Mexico City.

Though long out of print and unavailable except in libraries in Mexico, the *Diccionario de Geografia, Historia y Biografia,* by Alberto Leduc, Luis Lara y Pardo, and Carlos Roumagnac, published in France (in Spanish) in 1910, is a prolific source of information on old localities no longer on maps or the names of which have been changed. You will find many place names have been changed and frequently appear differently on different maps.

A new highway for west coast travelers has been opened recently between Mexicali and Santa Ana, Sonora. The new paved road, Federal Highway 2, connects with Federal Highway 15. The new 160-mile stretch saves 140 miles from southern California points.

Driving in Mexico

Watch the roads — some are good, others not as good. Be on guard for detours. Mexico is constantly building new roads and bridges and detours are common. Be especially careful when approaching bridges in Mexico. Many of them are quite narrow and dangerous. There is a custom frequently followed by Mexican drivers. If two cars approach a narrow bridge at the same time, the first one to flash its headlights has the right-of-way. It is sort of a game. In any case, if the car approaching you flashes its lights, don't try to beat him to the bridge, it might be a tie.

Frequently along the road you will observe the "Mexican sanitation department" in action. Sometimes horses, burros, cows, etc., are struck by autos or trucks and die along the road. There are usually several vultures around to pick the bones clean. Along the highway you frequently will pass a sign reading, "*Cuidado con el ganado,*" which means "Look out for the cattle." This is good advice.

In traffic, when the policeman faces you — stop; when he stands sideways — go. Mexicans with amazing bravado follow the deadly practice of whoever honks his horn first has the right-of-way. Don't be bluffed into giving an inch or you are lost! If you halve the speed limit in kilometers per hour, you will have the maximum speed in miles per hour with a margin of safety.

Be careful of the no parking signs in Mexico, particularly Mexico City. Cars that are illegally parked will have their license plates removed by the police. It costs about $4.00 in a time-consuming and roundabout process at a police station to get them back.

In many cities, the Mexican traffic police take your driver's license when issuing a ticket for traffic violations. They hold it as collateral until you appear at the police station on the charge. They give you a "receipt" which serves as a driver's license temporarily until you ransom your own.

The best gasoline to buy in Mexico is sold from the green and white pump and is called *Gasolmex.* The lowest octane gas is found in the red and white pump and is called *Supermexolina. Pemex-Sol* oil is a good grade of Pennsylvania oil. A satisfactory grade of Mexican oil is called *Mexolub.*

On entering Mexico from any point in the United States except on the east coast at Brownsville, you encounter a broad area of typical North American desert. Most of it is similar to the Sonoran and Mojave Deserts of the United States. In the west, it is actually an extension of these deserts. The same precautions taken in normal desert travel apply here, also. Be sure to carry water, digging tools, spare tires and reserve gasoline.

Trailering in Mexico

Using a trailer in Mexico can be quite enjoyable if travel is limited to the paved roads. Always drive with caution and watch for detours and road repairs. We broke our trailer hitch three times during the summer of 1960. Each break was due to the sudden appearance of road repairs.

We found it best when traveling off the paved highways to unhitch our trailer and leave it parked in the main square of the nearest town. We had no difficulty doing this on numerous occasions.

Communication

The language spoken is predominantly Spanish. It will help collecting and increase your pleasure if you know even a little Spanish. A Spanish dictionary and/or phrase book is helpful.

In asking directions, always inquire from more than one person. Never pose your question so that it can be answered with a simple yes or no, or the response invariably will be, *"Si, si, senor!"* The Mexican, by every instinct, is courteous, even to the extent of giving misinformation clothed in beguiling language if it is what he thinks you want to hear.

Keep all your faculties alert when listening to the two words *derecha* and *derecho*. The word ending with an "a" means to the right. The word ending with an "o" means you are correct, or straight ahead.

A hint on hints. Most Mexicans know (or think they know) the location of a mine. Guides are not always dependable and not all hints are 100 per cent accurate. Even professional mineral collectors spend three-quarters of their time tracing clues which usually turn out to be false. Be sure of your information or you will spend most of your days following erroneous leads.

It should be noted that the milages mentioned in this guide were taken on a speedometer without a one-tenth indicator. It is conceivable that they could be two miles off. It is always wise to ask directions and farther on to ask again.

Mexico is continually building new roads and with numerous detours, many milages indicated on maps may be inaccurate.

U.S. Border Regulations
Importation

The U.S. Customs is interested mainly in commercial quantities of ore that will be smelted, converted, or refined. Bringing in a few mineral specimens, provided you are not a commercial mineral collector, probably will entail only filing an informal entry permit; your minerals will be admitted duty-free. There are U.S. regulations covering just about all metals, in particular, copper, lead, zinc, silver, and gold. Lead- and zinc-bearing ores are subject to import quota regulations.

Specimens in the baggage of the arriving mineral or gem collector may be declared orally or, depending upon the value, in a written customs declaration. Specimens intended for sale or other commercial purpose, unless valued in excess of $250, may be entered on an informal entry. A resident returning from Mexico with dutiable specimens acquired as an incident of the journey, may include their value in any personal exemption. The exemption is not applicable, however, to any specimens collected on a journey taken principally for the purpose of collecting minerals and gems. When returning, a mineral and gem collector should show the customs inspector any receipt, bill, or invoice for items purchased. He may declare a fair market value for items acquired otherwise, i.e., by original discovery, barter or exchange, as a gift, etc.

Diamonds and other precious stones and gem minerals, rough or uncut, not advanced in condition or value from their natural state by cleaving, cutting, splitting, or other process, whether in their natural form or broken, and not set, are duty-free.

Diamonds, emeralds, rubies, sapphires, and other gem minerals cut but not set, suitable for use in the manufacture of jewelry, are dutiable at 10 per cent of the estimated valuation of the imported gems. Natural pearls which are not set or strung are dutiable at 10 per cent. Synthetic stones are dutiable at 10 per cent.

The duty on mounted jewelry varies between 30 and 55 per cent and should be checked at the border before entering Mexico.

The importation of gold into the United States is controlled by the *Gold Regulations* issued periodically by the United States Treasury Department and administered by the Office of Domestic Gold and Silver Operations. All gold must be declared at the point of entry. A person holding a Treasury Department industrial gold license may import gold in any form (other than gold coin) up to the amount authorized by his license.

A person regularly engaged in an industry, profession, or art requiring the use of gold may import not more than 50 fine troy ounces at any one time in any form (other than gold coin) without the necessity of obtaining a license, provided that the amount imported together with all gold which he has on hand at the time, does not exceed 50 fine troy ounces, and that the total acquired by him does not exceed 350 fine troy ounces in any one month.

Metals containing not more than five troy ounces of fine gold per short ton may be imported without a license, as may fabricated gold and gold in its natural state. Gold coins may be acquired outside the United States and imported only under a special license issued by the Office of Domestic Gold and Silver Operations, Treasury Department. Foreign gold coins minted before 5 April 1933, may be imported as rare, collector's items. Gold for sale to the United States Mints must be transported in customs custody to the mint. United States Mints will pay $35 (less ¼ of one per cent) per fine troy ounce, less any applicable mint charges for gold.

The preceding remarks on customs regulations are of a general nature and it is always a good policy to *check the latest regulations at the border* before entering Mexico. Any specific questions should be addressed to the nearest customs office.

Wrapping Delicate Specimens

Extremely fragile minerals such as aurichalcite, erythrite, etc., can be protected for transportation by first wrapping the edge with a thin, folded strip of newspaper. The delicate part of the specimen in the middle is then wrapped across the edge with additional newspaper. This method is used extensively by the people at Mapimí, Durango, and is highly recommended as a fast and efficient means of safeguarding delicate specimens in the field.

A Word to the Wise

Unless a mine is obviously abandoned, always obtain permission from the owner or manager before entering or collecting. Preferably, the visitor should be accompanied by a guide.

Most mines are dangerous. Always be on the lookout for abandoned shafts. There are many hazardous shafts in Mexico. They usually are not covered or filled. One false step could be your last.

One custom we especially liked in Mexico is that of waving gaily to all the people in the outlying areas as you drive by. Farmers working a quarter of a mile from the road will usually respond with warmth and a smile if they see you.

Let me stress again the advantage of learning and speaking Spanish. In finding your own localities, talk to the local inhabitants in Spanish and after you have gained their trust, they will generally guide you to the different mineral locations.

Remember, you represent not only yourself and your country, but all the gem and mineral collectors who come after you. Discourteous conduct on your part may reflect on others later. The Mexicans are a warm, friendly, and extremely proud people. Treat them with respect and you will be treated properly in return. *You* are a guest in *their* country.

Jade in Mexico

Genuine jadeite and nephrite artifacts have been found in Mexico and many of them can be seen in the museums. To date, however, there have been no authenticated reports of either jadeite or nephrite found *in situ*. There is a verified deposit in Guatemala, and perhaps the jade of the pre-Columbian civilizations came from there.

Since the earliest times, jade was honored and sought as a symbol of authority and wealth. However, the term was used very loosely and any green stone usually qualified as jade. Among the minerals that have been called "jade" are prehnite, green quartz (prase), turquoise, jadeite, nephrite, serpentine, and several common rocks. Still today, any green rock (including dyed travertine) qualifies as jade in the minds of many in Mexico. A great deal of "jade" is offered in shops and by street vendors. Little, if any, of it is genuine.

The Aztecs, particularly, valued jade above all other stones. The tributes of jade given to Cortes were valued more than gold and other objects. He received tributes from cities extending from the Guatemalan border northeast through the State of Chiapas, the Isthmus of Tehauntepec, and the four States of Oaxaca, Guerrero, Puebla, and Vera Cruz. If you want to search for jade *in situ*, this belt probably would be the likeliest prospect.

Recently a piece of blue gray material identified as nephrite by a gemologist, was displayed in Taxco. It is reported to have come from the mountains of Guerrero between Acapulco and the Rio Balsas. This report has not been authenticated, however.

Manto and Chimney Deposits

Two Mexican mining terms are frequently encountered, *manto* and *chimney*. The chimney is a pipelike replacement ore body of roughly circular cross-section in which the diameter is small in comparison with the length. The manto is a blanketlike, replacement ore body roughly parallel to the bedding in limestone (except in the Santa Eulalia District where they are like horizontal pipes). The *manto* (*blanket* in Spanish) is usually horizontal. The *chimney* is vertical or slightly inclined. Both of these ore bodies generally occur in limestone districts.

Some Mexican miners consider it bad luck for a woman to go underground in a mine. They believe the vein will play out. It is difficult or impossible in many areas for a woman to obtain permission to enter a mine.

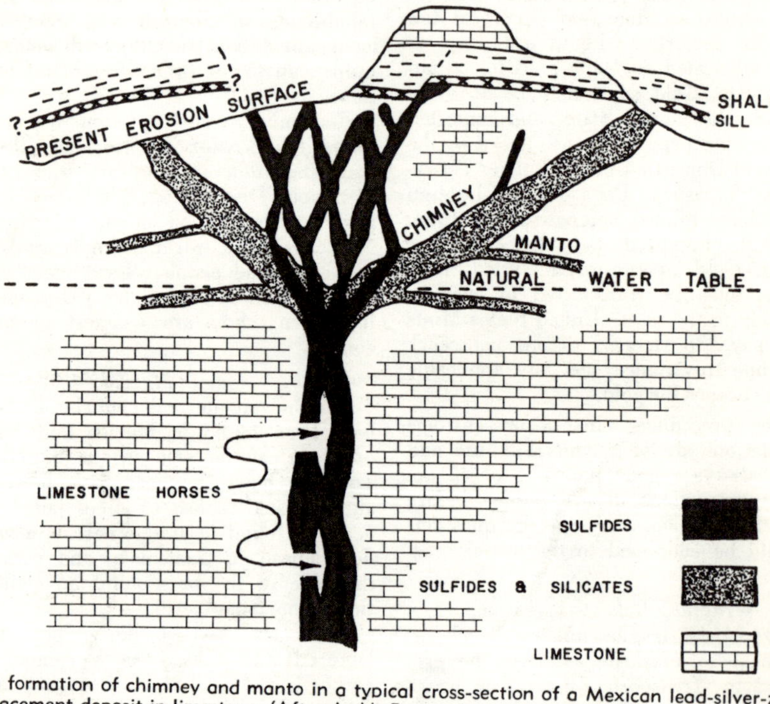

The formation of chimney and manto in a typical cross-section of a Mexican lead-silver-zinc replacement deposit in limestone. (After A. M. Bassett.)

MEXICAN AGATE

The most important agate-producing area in Mexico starts at Villa Ahumada in the State of Chihuahua, and extends for 70 miles in a southerly direction. It is also found in relatively smaller amounts in the State of Sonora. Farther south, in the State of Durango, agate is found near Bolson de Mapimí, Rodeo, and probably near Coneto de Comonfort.

The agates in northern Mexico were not exploited commercially until after World War II. The first Mexican agate to be collected for export is believed to have come from the Gallego and Sueco ranches on Highway 45 between Juarez and the City of Chihuahua. It was highly weathered but quite colorful.

Most of the good agate deposits are on large, private cattle ranches and visitors are not allowed to collect. Most of the surface agate has been recovered from these ranches the majority of which are in rough country devoid of roads. Agate *in situ* can still be found, but it is improbable that the owners will allow extensive mining to disrupt their cattle business. The geology is, in general, older volcanics, including some altered rhyolites. Some of the ranches have chalcedony replacements of barite and aragonite crystals.

Of all the nodules found, only about ten per cent are two or more inches in diameter, which would be considered medium sized. Gem grade nodules up to 70 pounds have been found. About ten small nodules are recovered for each

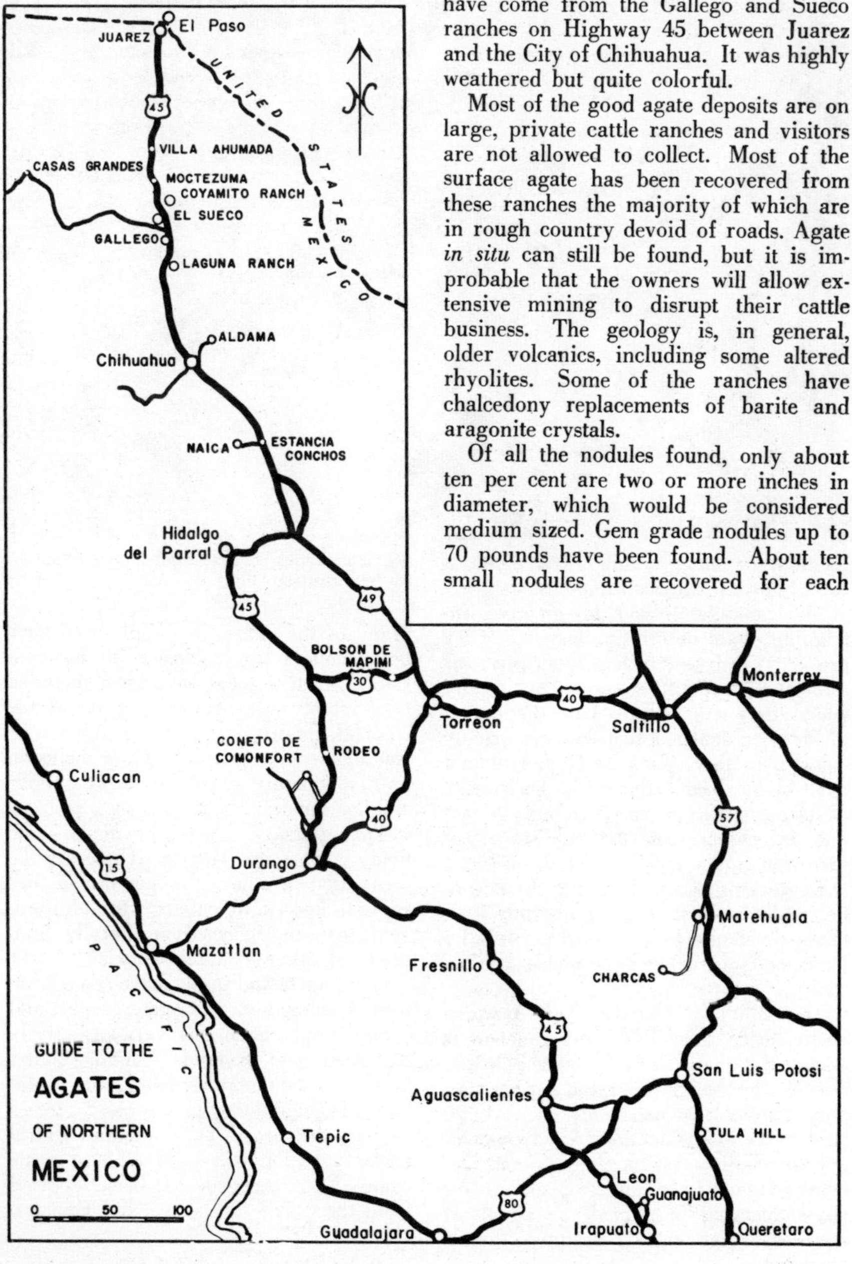

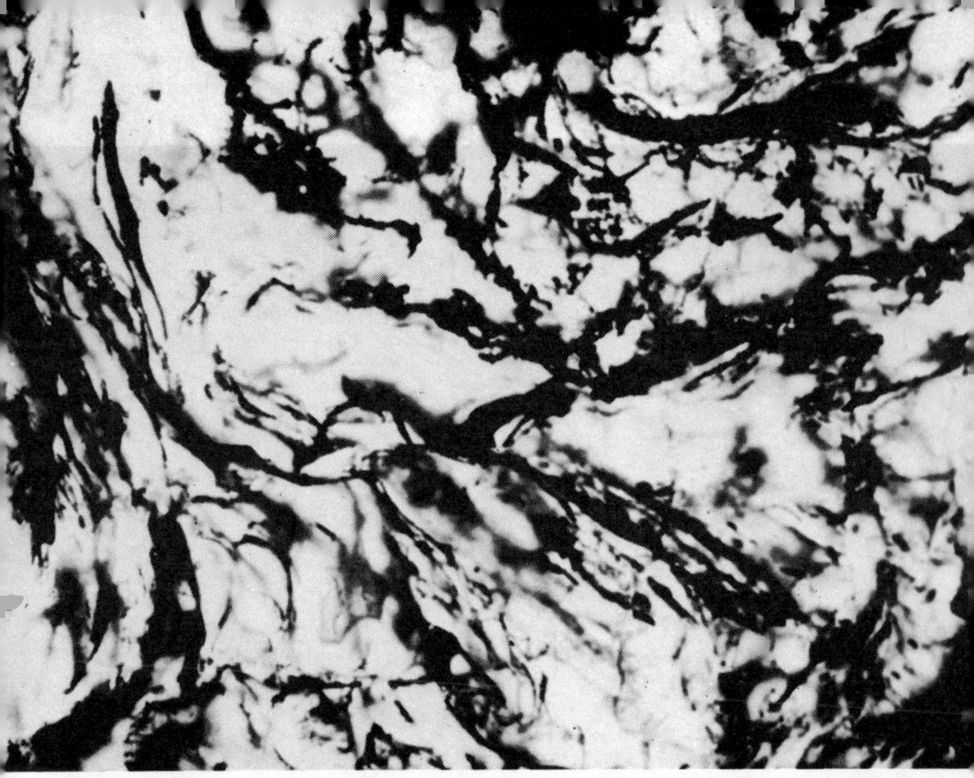

medium or large nodule. Iris agate is found in some Mexican nodules.

Mexican agates have imaginative, intriguing, and descriptive names. They are of two types, either descriptive or after the name of the ranch or locality where they are found. One distinction of Mexican agates is that they are usually colorful in the natural state and do not have to be dyed artificially. The agates of northern Mexico are frequently *tested*, that is, chipped on the side to make sure that agate will be found inside.

In working agate, dynamite should not be employed as it causes numerous fractures. It should be extracted entirely by hand methods, using bars, picks, wedges, etc.

The following glossary of Mexican agate terms, localities, and ranches is arranged alphabetically by principal category. The range of retail prices per pound should be used only as a rough guide. The prices fluctuate widely according to quality, rarity, supply, and demand. Agates have numerous colors and no attempt has been made to list all of the colors for each descriptive name.

Mexican black plume agate photographed by transmitted light.

Many of the names in the glossary have been thought up by dealers to promote sales. Most of them have been included here merely as a guide to some of the colorful, descriptive — but at times misleading — nomenclature. Their inclusion in this glossary does not imply mineralogical sanction.

This glossary covers the main agate fields in northern Mexico. These are the areas that produce the popular agates so well known to cutters. In addition, agate can be collected in literally hundreds of other places in Mexico.

Good agate and jasper have come from near Alamos, Sonora. Agate, jasper, and common opal, much of it very attractively patterned and colored, can be found in the mountain areas south and southeast of Querétaro. Nearly any stream bed in the broad band of the central volcanic region will yield at least some agate. Excellent agate has been reported from the States of Oaxaca and Chiapas, also.

Glossary of Mexican Agate Terms

Aldama, a town in the State of Chihuahua

amethystine (amethyst). Purple agate, usually in white chert, sometimes with banded agate; $1.00 to $1.50.

Apache agate. A nodular agate predominantly bright red and vivid orange but sometimes with light to dark blue, yellow, tan and white; $1.00 to $5.00.

Apache, Chihuahua. A small village about 50 miles northwest of Casas Grandes.

bean agate. Occurs as small, elongated nodules from about one-quarter to two inches in length. It is found south of Villa Ahumada and on the Moctezuma Ranch; $1.25 to $1.50.

bird-of-paradise agate (sometimes known as flower-of-paradise agate). Has plume feathery red formations in an opaque to translucent matrix; other colors include brown, cream, yellow, pink, purple, tan, white, and green; $0.75 to $2.00.

black and white agate. A banded agate; $1.00.

blue agates. Banded nodules associated with agates of other colors. Found west of Villa Ahumada; $1.25 to $2.00.

blue shadow agate. Nodular; $1.00.

bouquet or flowering agate. A type of plume agate.

carnelian. A transparent to translucent red chalcedony.

Casas Grandes, a town in the State of Chihuahua.

Casas Grandes agate. A nodular gray, red, pale pink to purple agate, sometimes showing good fortification; $1.25 to $2.25.

cathedral agate. Has a cathedral-like shape of red agate overlain by white to gray agate. It also has several other colors; $3.00.

cedillos. Colorless, gray to blue vein or nodular agate; $1.25.

Christmas agate. A moss agate with long moss formations. It is sometimes listed as a flame agate as it sometimes has red and green "flames" in a clear vein agate; $1.00 to $1.50.

confetti agate. A type of bird-of-paradise agate from the Naica area, State of Chihuahua. Colors include reds, pink, and yellow; $0.75 to $1.00.

Coyamito agate. Considered by many to be the best found in Mexico. Very little of this agate is being produced at this time as most of the surface material has been collected and the owner is not interested in having his cattle ranch dug up. This nodular agate sometimes has eye patterns and tube-like formations. Most of the nodules contain hollow centers. Pseudomorphs after aragonite are found in the Coyamito Ranch area. The colors include gray, green, tan, white, light red, yellow, pink and purple. Priced generally from $1.00 to $3.00.

Coyamito Ranch. A ranch in the State of Chihuahua, southeast of Moctezuma.

crazy lace agate. See lace agate.

cyclops agate. Red and white nodular agate exhibiting a single "eye." From the State of Chihuahua.

dendritic agate. See moss agate.

docina agate. Black to white to clear agate, topped with quartz crystals. Fluorescent. $1.50.

eye agate. Nodular agate with a solid color "eye" in the middle; $1.50 to $3.00.

fire agate. Composed of botryoidal chalcedony with a brownish sheen and play of colors due to inclusions of "limonite." It is reported to be found in gem grade at Tula Hill, 45 miles south of the city of San Luis Potosi, on the east side of Highway 57 between kilometers 359 and 361. This area is filed upon by Esteban Bernal of Charcas, State of San Luis Potosi. Usually less than five per cent of the material mined is of gem grade; $1.50 to $5.00.

fisheye agate. A nodular or vein agate usually composed of white dots in a pink or blue material. Considered to be a type of eye or flame agate. Found in the Apache area.

flame agate. A vein material with red "flames" in clear agate. The name is sometimes applied to a low-grade moss agate. Also considered to be a type of Apache agate from Apache. Names include: Mexican red flame, red flame, Christmas, and forest fire flame agate. Forest fire flame agate has black flames topped with red. Some flame agate has been found west of Villa Ahumada, Chihuahua; $0.80 to $1.50.

fortification agate. Usually nodular agate with angular bands in the shape of a fortress.

Gallego agate. Considered by many to be some of the finest banded agate found in Mexico. Many colors,. including red and gray fortification nodular agate, are found. Some iris agate also has been reported from an altered rhyolite about seven or eight miles east of the highway on the Gallego Ranch. One authority has estimated that about 50 per cent of the large nodules are hollow and 20 per cent of the smaller nodules have cavities. A rich purple banded agate is found east of Gallego. Nodules $3.00.

Gallego Ranch. A ranch in the State of Chihuahua. Center of the gem grade banded agate belt of northern Mexico.

high fidelity agate (a proprietary name). Probably found near Coneto de Comonfort in the State of Durango. This agate is usually found in veins and occasionally in nodular form. The colors include red, lavender, blue, yellow, rose, purple, black, and white. Generally there are two or more colors per stone; $1.60 to $2.50.

iris agate. The rainbow colors in iris agate are believed to be due to the interference

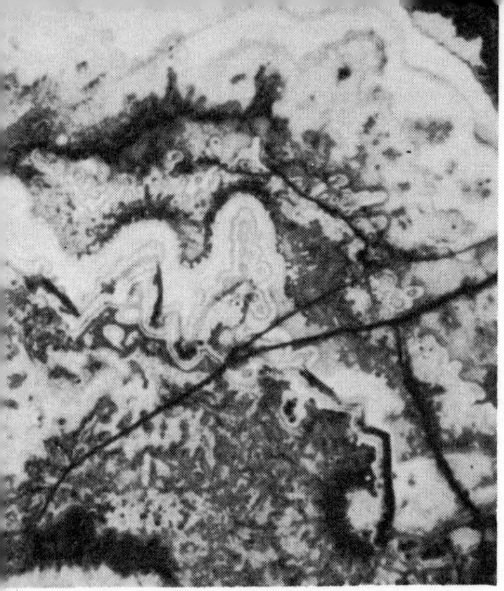

Crazy lace agate, Mexico.

of light passing through closely spaced bands. The color could come from a difference in refractive index between the bands which are composed of quartz and water-filled micropores. In the past small amounts of iris agate have been found on the Sueco Ranch. At the present most of it is found on the adjoining Gallego Ranch.

lace agate. Generally a vein, but also a nodular agate. Colors: gray, white, black, tan, red, and pink. Names include black lace, black and white lace, crazy lace, Mexican lace, white lace, and purple lace agate. The popular crazy lace agate is bright red and orange. It is reported that lace agate can be found west of Villa Ahumada; $0.50 to $2.25.

Laguna agate. Also considered by many to be the best grade found in Mexico. Nodular and vein agates are found on the Laguna Ranch. This ranch has few hollow nodules and less crystalline quartz than other Mexican nodular agate. The Laguna Ranch also produces red and blue flame agate, fortification agate and eye agate. Colors are varied and include blue, yellow, lavender, deep red, wine-red, purple, and pink. Small nodules $1.25. Medium and large banded nodules $3.00 to $4.50.

Laguna Ranch. A ranch in the State of Chihuahua. Also known as the Borunda Ranch or Ojo Laguna. Owned by Enrique Borunda. It is about 20 miles south of Gallego, Chihuahua. According to the owner, the agate from this ranch is sold exclusively to the Southern Gem Mining Company of El Paso, Texas. The owner reports that no one has ever been given permission to enter the Laguna Ranch for the purpose of collecting or purchasing agate.

Loma Pinta (Painted Hill) agate. Nodular agate from the Painted Hills in the eastern part of the Moctezuma Ranch.

luna (moon) agate. A translucent to opaque agate, amethyst in color, with moonlike white dots; $1.40 to $3.75.

Moctezuma. A town in the State of Chihuahua.

Moctezuma agate. Beautiful, circular nodules with banded fortification agate. Colors: yellow, tan, pink, blue, and rarely green or red. Loma Pinta agate is a type of Moctezuma agate; $1.00 to $2.75.

moon agate. See luna agate.

moss agate. Also known as dendritic agate, has dendritic or mosslike inclusions of chlorite, iron or manganese oxides, etc. Red and yellow moss agate is found in the Naica area, State of Chihuahua. An attractive white opaque agate with black dendrites is found in the State of Durango; $0.60 to $1.00.

paradise agate. See bird-of-paradise agate

Parcellas agate. Found near Parcellas, about 40 miles east of Casas Grandes. Pink or lavender nodules $1.50.

plume agate. Featherlike inclusions in transparent to opaque chalcedony. Some of these inclusions are as much as 14 inches long. The colored inclusions are black, red, white, or golden. Names include black plume, red plume, etc. Red plume agate has been found southeast of Reforma, State of Chihuahua. Some black plume agate is fluorescent under short wave.

royal purple Aztec agate, also known as rodeo royal Aztec agate, royal Aztec agate, and royal Aztec purple banded agate (a proprietary name). Vein and nodular agate found on and to the east of the Gallego Ranch. Also found near Rodeo, State of Durango. Primarily a vein agate, although some pieces contain beautiful flowery formations. This material is composed of purple and yellow or clear bands of agate; $0.75 to $2.00.

sagenite agate. Red and yellow bursts (jasper); $1.50.

Saint (Santa) Gertrude moss agate. A brown and yellow agate found in southern Chihuahua; $1.00.

satellite agate (a proprietary name). Veinlike pieces of botryoidal and stalactitic forms. From the Durango area. $2.80.

splash agate. A red vein moss agate, sometimes called red splash agate.

Sueco agate. This agate is similar to that found at Gallego Ranch. Less iris agate is found and an estimated 80 per cent of the nodules are hollow or have centers of quartz. Small banded nodules are found; $1.25.

Sueco (Swedish) Ranch. A ranch in the State of Chihuahua.

Villa Ahumada, a town in the State of Chihuahua. Banded agate is found southeast of here.

SECTION 1

Localities in this section are generally accessible from Federal Highway 15. This is the main west coast route from Nogales to Guadalajara. The localities are arranged as nearly as possible from north to south.

Many cities and states in Mexico have the same names. You will also find many duplicate names in different, or even the same states. Over the years place names have been changed, often several times. We have tried, as far as possible, to use current names. Approximate locations of the areas covered are given on the Index Map on pages 46 and 47.

References such as Chihuahua, Chihuahua, or Durango, Durango, mean the first is the city, the second the state. Where confusion might result whether a single name is a city or state, it is always the city unless followed by a small (s). Durango means City of Durango; Durango(s) means State of Durango.

El Antimonio District

The Antimonio District, in northwestern Sonora(s), is about 99.4 miles by road west of Santa Ana and 62 miles south of Ajo, Arizona. To reach this District from Santa Ana you pass through the towns of Altar and Caborca. The District was known as early as 1722. It is the only District in North America that normally produces antimony ore from placers. The antimony deposits, both placer and vein, are distributed throughout an area 4.7 miles long from north to south and 0.6-2 miles wide.

The District has 22 mines working the largest and richest deposits. The main mines are owned by Roberto Moreno with smaller deposits owned by Rafael Ruiz and Roberto Perez. Total production has been about 30,000 tons of metallic antimony.

The bedrock of the District consists largely of sandstone, siltstone and conglomerate overlying limestone. Thrust faults are common and are occupied by veins containing quartz, oxides of antimony, and native gold with silver chlorides and bromides. The silver is always associated with the antimony oxides, while the gold is enclosed in the quartz. The gangue is milky quartz or finely grained chalcedony. The veins average 12 inches wide by 300 feet long. The maximum width is 10 feet, the maximum length 1,640 feet.

When a vein is discovered it is mined from the surface to a depth of less than 30 feet, then developed to greater depth by sinking vertical shafts. Usually the vein is mined upward. Many of the shafts along the vein are only 20 to 30 feet apart. All the mining is done by hand methods and except at the four principal mines in the District, the ore and waste rock are removed from the vertical shafts by hand hoists.

The Fortuna Mine in this District is the locality where the mineral byströmite was first found.

El Antimonio District Mines

Argentina, La	Palo Verde, El
Carta Blanca, La	Piedra Azul, La
Divisadero, El	Pensamiento, El
Fortuna, La	Promontorio, El
Limena, La	San Francisco
Loca, La	San José
Montaña, La	San Miguel

Minerals Reported

Antimony oxides	Quartz
Bromyrite	Quartz, var. Chalcedony
Bystromite	Stibiconite
Calcite	Stibnite
Cerargyrite	
Gold, Native	

CANANEA DISTRICT

The Cananea Copper District in Sonora(s), is in the Cananea Range, just south of the Mexican border, between Nogales and Naco, Arizona. Mining commenced before 1865. The District is first in total copper production in Mexico. It has yielded over 1.5 billion pounds of copper, 25 million ounces of silver, and total minerals exceeding 230 million dollars in value.

The District consists mainly of porphyry copper deposits, breccia pipes, minor limestone replacements, and contact-metasomatic deposits. The geological history of the District is quite complex.

The principal ore minerals are bornite, chalcopyrite and molybdenite with minor amounts of pyrite, tennantite, covellite, and chalcocite. The most abundant ore mineral is chalcopyrite.

The Colorada Mine near the head of Democrata gulch on the northeast slope of the Cananea Range, contains an outstanding feature in the rich *Colorada* ore pipe. This ore body lies some seven or eight hundred feet beneath the present erosion surface, and is shaped like a goblet.

Of unusual mineralogical and gemological interest is the transparent facet grade sphalerite found in the District. The sphalerite is a yellow olive-green, and brown to red. A pale green is considered the most desirable. Transparent green sphalerite has been found at the Manzanal Mine and a golden sphalerite at the Chivera Mine.

Cananea District Mines

Chivera	Manzanal
Colorada	Veta
Duluth	

Minerals Reported

Alunite	Germanite (?)
Amphibole group	Hedenbergite
Andradite	Idocrase (Vesuvianite) (?)
Azurite	Kaolinite (?)
Barite	"Limonite"
Bornite	Malachite
Calcite	Molybdenite
Chalcocite	Muscovite, var. Sericite
Chalcopyrite	Paratellurite
Chlorite	Phlogopite
Chrysocolla	Pyrite
Cobaltite	Pyrrhotite
Copper, Native	Quartz
Covellite	Sphalerite
Cuprite	Tellurite
Digenite	Tellurium, Native
Dolomite	Tennantite
Enargite	Tetrahedrite
Epidote	Tetrahedrite, var. Freibergite
Erythrite	Zoisite (?)
Famatinite (?)	
Feldspar group	
Galena	

Mexico is full of attractive sights like this view of the cathedral at Guaymas with the Gulf of California in the background.

ARIZPE DISTRICT

Many mineral collections throughout the world contain fine silver specimens from Las Chispas Mine near Arizpé, Sonora. Mining here, in a country rock of rhyolite, began in 1880. The ore averaged as high as 2,000 ounces of silver per ton. An 82-pound mass of polybasite in the U.S. National Museum from there assayed 16,868 ounces of silver per ton. Unusually fine crystals of stephanite and polybasite are found at Las Chispas Mine. Small (up to 1/16 inch), beautiful fluorite crystals occur on the silver minerals, especially polybasite. The Pedrazzini and San Lorenzo Mines are also in this District.

Minerals Reported

Argentite	Polybasite
Calcite	Pyrargyrite
Chalcopyrite	Pyrite
Cuprotungstite	Quartz
Fluorite	Silver, Native
Gold, Native	Stephanite

NACOZARI DISTRICT

The Nacozari District in Sonora(s), ranks third in total copper production in Mexico. The Pilares Mine alone has yielded over 600 million pounds of copper and four million ounces of silver. It is located in a volcanic region. Hot water solutions filled the altered breccia spaces in the volcanics with ore minerals, principally iron and copper sulfides, hematite, var. specularite, and scheelite. Chalcocite enrichment has occurred to a depth of 200 feet. The ore averages two and one-half to three per cent of copper.

Pilares de Nacozari Mine and Las Pilares Mine are also in this District.

Minerals Reported

Bornite	Pyrite
Cerussite	Quartz
Chalcocite	Rutile
Chalcopyrite	Scheelite
Davidite	Stibiobismuthinite
Dumortierite	Torbernite
Hematite, var. Specularite	

MOCTEZUMA DISTRICT

The Moctezuma Mining District is in Sonora(s). (It should not be confused with Moctezuma, Chihuahua, the popular agate area. See *Agates.*)

All of the mines in the Moctezuma area are on private property and are either being developed or are otherwise inaccessible to collectors. Mines here include Candelaria, Guadalupe, and the Moctezuma.

Minerals Reported

Antimony, Native	Paratellurite
Barite	Pyrite
Chalcostibite	Scheelite
Denningite	Selenium, Native
Emmonsite	Spiroffite
Gold, Native	Tellurite
Illite	Tellurium, Native
Jarosite	Tellur-Jarosite
Klockmannite	Tetrahedrite
Mackayite	

HERMOSILLO DISTRICT

Hermosillo (air-mo-SEE-yo) which means, "little beauty," is the capital of the State of Sonora. Engineer Ramón Fernández is in charge of the local Comisión de Fomento Minero at Hidalgo No. 74.

The Hermosillo District is a low-grade contact metamorphic tungsten deposit. It was discovered in January 1943.

Minerals Reported

Calcite	Powellite
Chalcopyrite	Pyrite
Epidote	Quartz
Fluorite	Scheelite
Garnet group	

ALAMOS DISTRICT

Turning off the main highway at Navojoa, 32 miles east is Alamos (AH-lah-mos), Sonora, a favorite retreat of Mexicans and Americans. It is a quaint village with cobblestone streets. There is a small park in the center of town with beautiful, tall, feather palms, and the ever present bandstand in the center.

Coronado and his men camped in Alamos in 1531. In 1781, it was known as the silver capital of the world, and had a population of 30,000.

Alamos is now protected by Mexico as a Government Colonial Monument. The lovely exteriors of the buildings with their Moorish arches, muted golden colors and fretted casements, are safe from the plague of modernization. Today, only the interiors of buildings may be remodeled.

Alamos is the former home of Alberto E. Maas (Don Alberto) who dealt in minerals for years. He first found and sold "peanut rock," an obsidian with red orbicular inclusions. It has also been called "almasite," but the name is not recognized officially. Unusual quartz specimens with a radial structure have been found near Alamos. Quartz crystals covered with epidote are found in the area.

A reliable guide to the mines is Sr. Fernando Peralta, No. 26 Calle Juarez. He used to ride with Pancho Villa and is quite familiar with the local terrain. His fee is about $4.00 per day.

Copper Mines

Thirteen miles east of Alamos, over a rough road, are four copper mines. From north to south along the strike of the intrusion: Empaco, La Aztecita, La Azteca, and Malaquita. The copper occurs in a dike three to eight feet wide and about two miles long. The copper ore is low-grade malachite, azurite, and chrysocolla. It is too low-grade to make mining profitable, at present.

Nice specimens of malachite on drusy quartz can be collected at the Malaquita Mine. A sledge hammer is useful at this locality. The most interesting specimens from the Azteca Mine are small crystals of dioptase on quartz. These specimens make quite attractive micromounts.

Gnats in this area are bothersome and an insect repellent is desirable. A guide from Alamos is recommended for this area, as you could easily get lost.

Minerals Reported

Alamosite	Glaucodot
Andradite	Grossularite
Azurite	Malachite
Calcite	Pyrolusite
Chrysocolla	Quartz
Clinozoisite	Scheelite
Cobaltite	Stilbite
Conichalcite	Torbernite
Dioptase	Tourmaline, var. Dravite
Epidote	
Erythrite	Tungstite

San Bernardo Area

In the San Bernardo area about 27 miles north of Alamos, is the Sara Alicia Mine where erythrite can be collected. Collecting erythrite is laborious and sometimes requires dynamite. To visit this mine, a guide is necessary. Plan on a two- or three-day trip, as you must travel to the mine by horseback.

Minerals Reported

Arsenopyrite	Harmotome
Cobaltite	Loellingite
Erythrite	Pyrolusite
Glaucodot	Scorodite

If you buy or collect erythrite you will need metal cans and wax to protect the specimens. Erythrite is very fragile. Just touching it will cause the delicate crystals to crumble. Heat some wax in a tin can and imbed just the lower part of the matrix. This will protect the specimen on the trip home.

ALISTOS DISTRICT

The Alistos District, Sinaloa, can be reached from Highway 15 by taking the road east from Guamuchil to Mororito, thence southeast to Badiraguato, and finally northeast to Alistos (also spelled Alisos).

The mineral association here of native gold with niccolite and millerite is considered unusual. The gold-nickel ore of this area is of two types:

1. Gold-massive niccolite containing minor amounts of annabergite, gersdorffite, jarosite, "limonite," maucherite, etc. Niccolite is the most abundant metallic mineral in this ore.
2. Gold-millerite with associated erythrite, gersdorffite, jarosite, "limonite," pentlandite, pyrite, and violarite.

Millerite, pentlandite, and violarite do not occur in the gold-niccolite type of ore. Niccolite and maucherite are not present in the ore of the gold-millerite type. No pyrrhotite was found in either type of ore. The mineralization responsible for the unique mineral associations is believed to be of hypogene (ascending waters) origin.

At the Gloria property, Alistos, the mineral assemblages of the ore are extraordinary, particularly native gold with niccolite and primary millerite also containing gold. The rare minerals of the Alistos area include annabergite, gersdorffite, maucherite, and violarite.

Minerals Reported

Actinolite	"Limonite"
Annabergite	Maucherite
Calcite	Millerite
Chlorite	Niccolite
Erythrite	Pentlandite
Gersdorffite	Pyrite
Gold, Native	Pyrrhotite
Ilvaite	Siderite
Jarosite	Violarite

A four-wheel drive vehicle, like this Travel-all is an ideal vehicle for getting into areas in the back country of Sonora (s).

MAGDALENA

Some of Mexico's finest opal is mined near Magdalena, Jalisco. This village is about 50 miles northwest of Guadalajara on Highway 15. It is close to Tequila, home of Mexico's fiery drink. The opal is similar to that from the Querétaro area and buyers from this area regularly visit the village. Much opal sold in Querétaro is mined near Magdalena.

This area has not been scouted personally but the following, taken from two individual reports, is believed to be reliable:

Sr. Arturo Salas, Motel and Restaurant Avenida, Magdalena, Jal., has stated, "We have the so called rain garden (fire), harlequin, white opal, opal with orange and yellow, and blue, purple and green. . .

"There are in this region some other types of gem rocks and minerals . . . such as amethyst, topaz, and gold and silver mines . . . we have optical calcite. . .

"The largest mine near Magdalena is the San Simón, about four miles west on the main highway, then right (north) another three miles to the village of San Simón . . . in this mine they often find harlequin and rain fire types. . . .

"Another mine is La Amazata, a group of beds above the little village of La Amazata. It is about two miles south of Magdalena by automobile and then a walk or horseback ride of two miles to the mine. . . .

"Another mine is Cinco Minas. This mine is above Hosto, about four miles

The dike which carries the copper at the Azteca mine is from three to eight feet wide and over two miles long.

from Magdalena. It is famous because of gold and silver. . . .

"These three mines produce opal of the finest quality. Other mines (in the area) include the Santo Domingo, La Yesca, El Favor, El Lorito, La Calabaza, Las Latillas, Oconahua, San Andres, and Las Barrancas."

Sr. Salas reports that guides are available in Magdalena and collecting is allowed in some of the mines.

An American collector who visited the area in 1962 reported:

"We learned of the following mines: La Unica, Rancho Huizilap; La Latilla; Barranquitas; La Estancia, 10 km out; San Simon; and La Quireda, San Andres, 15 km.

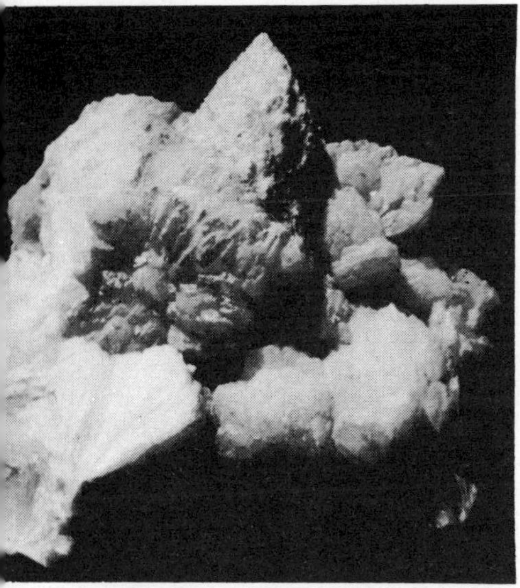

Stilbite from near Alamos, Sonora, Mexico.

"The countryside between Tequila and Magdalena is literally strewn with obsidian, veins two to three feet thick in the road cuts, small pebbles imbedded in adobe walls. Large chunks are in stone walls . . . and even in concrete aggregate in the light standards of Magdalena. We hacked away in one area but found none with sheen. We did break some that showed cloudy patterns. [Kunz reported sheen obsidian from this area in 1902.]

"We also found a conglomerate formation of concretions of soft clay balls with blue centers. They probably originated from mineral hot springs or mud

The main highway cuts through a huge lava flow 38 miles south of Tepic.

pots, but it was among the obsidian boulders.

"Saturdays and Sundays, opal miners come to Magdalena to sell their opals. Mines are accessible by Jeep or truck only 10 kms (about 7 miles) from Magdalena to San Andres and from there to La Estancia opal mine."

This area is relatively unpublicized even though it is on the main coast highway from Nogales to Guadalajara. It is the reporter's opinion that the area offers much of interest to collectors. He suggested Sr. Ladislao Sambrano, Calle Tabasco No. 37, Tequila, Jalisco, as a good guide to the area.

The famous tequila distilleries in Tequila are recommended for a visit.

GUADALAJARA

Guadalajara (gwah-dah-lah-HAH-rah) is the capital of the State of Jalisco (ha-LEES-co). It is the second largest city in Mexico, with a population of 400,000. Modernization is rapidly taking place in Guadalajara.

Engineer Arturo Abril, Terminal Autotransportes, 60 Piso Desp. 43, is in charge of the local office of the Comisión de Fomento Minero.

Much of the road you will ride over in the area comes from this quarry, 12 miles south of Tepic.

SECTION 2

Districts and areas in this section generally are accessible from Federal Highway 45, the central route from Ciudad Juarez to Mexico City. Localities are listed from north to south in so far as possible. The main agate areas are also along this route. See "Agate" for details of agate localities.

SABINAL DISTRICT

The Sabinal District is 6.8 miles west of the Sabinal station on the Mexican Northwestern Railroad, Municipio of La Ascensión, Chihuahua. Two types of deposits occur in the District:

1. Pyrometasomatic deposits along the contact of the igneous intrusion and the sediments.
2. Fissure vein deposits in limestone.

The bulk of the silver production of the district has come from the Florencia-Magda vein. This district has been mined since 1821.

Native silver is one of the most important of the metallic minerals in the deposit and is associated with sphalerite, galena, barite, quartz, calcite and rarely, argentite. The native silver is of interest because it was deposited from primary or hypogene solutions. The gangue minerals found in the veins include barite, calcite, quartz, siderite, fluorite, and iron oxides. In addition, pseudomorphs of quartz after barite and calcite are common.

The Adventurera mine is found in this District.

Minerals Reported

Acanthite	"Limonite"
Albite	Malachite
Anglesite	Polybasite
Argentite	Proustite
Barite	Pyrargyrite
Calcite	Pyrite
Cerargyrite	Quartz
Chalcopyrite	Rammelsbergite
Chrysocolla	Safflorite
Covellite	Siderite
Fluorite	Silver, Native
Galena	Sphalerite
Garnierite (?)	Tetrahedrite
Gold, Native	

LOS LAMENTOS DISTRICT

Los Lamentos District is in the Sierra de los Lamentos, Municipio of Villa Ahuamada, Chihuahua. It was discovered in 1923. The area can be reached by road and is 40 miles east of Villa Ahumada on Highway 45. As is true of most silver, lead, and zinc districts in Mexico, Los Lamentos has suffered from the low price of metals, and water in the bottom of the mines. Erupcion and Ahumada are the two main mines of the District.

The ore bodies at Las Lamentos are of the chimney type and are found in limestone. They are composed largely of oxidized lead, zinc, and iron minerals. The chief ore minerals are anglesite and cerussite. Cerussite is the principal ore mineral in the lower levels. The gangue minerals include calcite, dolomite, gypsum, and sulfur.

This District long has been known to mineral collectors as the type locality of orange wulfenite crystals. These richly colored, square, tabular crystals are found abundantly in the Ahumada Mine but rarely in the Erupcion Mine.

Minerals Reported

Anglesite	Hydrozincite
Aragonite	Jarosite
Brochantite	Plumbojarosite
Calcite	Pyrite
Cerussite	Sphalerite
Descloizite	Sulfur
Dolomite	Vanadinite
Galena	Vanadinite, var. Endlichite
Goethite	Willemite
Gypsum	Wulfenite
Hematite	

PLACER DE GUADALUPE PUERTO DEL AIRE DISTRICT

The Placer de Guadalupe-Puerto del Aire District is in the Municipio of Aldama, Chihuahua. Placer de Guadalupe is about 60 miles northeast of the City of Chihuahua. Puerto del Aire is two miles east of Placer de Guadalupe. The country rock is black shale interbedded with sandstone and limestone.

This District is noted among mineralogists for the rather unusual mineral association of native gold with uraninite (pitchblende). The uraninite and native gold are found in a calcite vein. The veins vary in width from eight inches to one foot.

The uraninite occurs as crystals or aggregates of crystals ranging up to an inch in diameter. The crystals are commonly cubic or octahedral. A greater amount of uraninite has been found at Puerto del Aire. Native gold also can be found in this District without associated uraninite. It is found in quartz and placers.

Placer de Guadalupe-Puerto del Aire District Mines:

Bolanos	San Blas
Chiva, La	San Martín
Judia, La	San Rafael
Puerto del Aire	Virgen, La
Purísima, La	

Minerals Reported

Albite	Malachite
Beryl	Muscovite, var. Sericite
Biotite	Orthoclase
Calcite	Pyrite
Carnotite	Pyrolusite
Chalcopyrite	Quartz
Chlorite	Titanite
Fergusonite	Uraninite (Pitchblende)
Gold, Native	Uranophane
Kaolinite	
"Limonite"	
Magnetite	

CHIHUAHUA

Chihuahua (chee-WAH-wah) is the capital of Mexico's largest state of the same name. Senior Engineer Carlos García is the director of the Comisión de Fomento Minero in Chihuahua, at San Martín No. 502.

There is a mineral dealer operating under the name of Gemexa, at Paseo Bolívar No. 206. He sells minerals from Mexico and has an excellent selection of tumbled stones. Selenite crystals can be purchased for 80 cents per pound and calcite for 70 cents per pound. Directly across the street from Gemexa is a Museum of Minerals specializing in minerals from the State of Chihuahua, and Mexico in general.

SANTA EULALIA DISTRICT

The Santa Eulalia District is second in terms of total silver production in Mexico. The town of Santa Eulalia is one of the oldest in northern Mexico. This District has produced over 300 million dollars' worth. The District was discovered about 1591.

The turn-off is about seven miles south of Chihuahua on Highway 45. It is in the Municipio of Aquiles Serdán, Chihuahua(s).

This District produces mainly silver, lead, and zinc. The productive area has been divided into the western and eastern camps separated by three miles of apparently barren ground. The District has more than 500 miles of underground workings. Most of it is owned by the Howe Sound Company. The general geology consists of limestone, argillaceous material, limestone conglomerates, volcanic flows, tuffs and marls.

An interesting alteration product of galena has been found here. It consists of banded nodules of anglesite and cerussite with galena cores.

The rare minerals reported include alunogen, fayalite, goslarite, ilvaite, iodyrite, and knebelite.

El Potosí Mine

One of the most productive mines in the District is the El Potosí. It has been worked since 1703 as a rich silver-lead producer. The mine is owned by the Howe Sound Company. It is second only to the Fresnillo Mine, Fresnillo, Zacatecas, in terms of total silver production. It has an estimated 250 miles of underground tunnels which disclose mantos, chimneys, and irregular replacements.

The Potosí Mine has a high-grade sulfide ore, low-grade sulfide-silicate ore, and oxidized ore which is now largely exhausted. The unusual silicate ore is composed of actinolite, fayalite, hedenbergite, hematite, ilvaite, magnetite, quartz, and silicified limestone.

Beautiful specimens of quartz stained by chrysocolla have been found in this mine.

San Antonio Mine

The San Antonio Mine is 4.3 miles from Santa Eulalia. It has produced silver, lead, tin, vanadium, and zinc. The principal by-products are copper and gold. The lead-zinc-silver ores were described by Prescott as replacement deposits in limestone. The chimney type of ore deposit here is found in a thick series of Cretaceous limestones.

Geologically the area around the San Antonio Mine is composed of Cretaceous limestone overlain by beds of Tertiary clastic and volcanic rocks.

This unique mine is noted as the first successful tin producer in North America. Tin was discovered here in 1926. From 1926 to 1939, about 3,000 tons of tin were produced. In 1940, this was half of Mexico's total tin production. Tin (cassiterite) in association with an ore deposit is an unusual occurrence.

Usually the tin is found in veins or the placers derived from them.

Quartz crystals encrusted with pyrite from this mine are very attractive.

Santa Eulalia District Mines

Abundancia
Aguado
Ampliacion
Bolivia
Buena Tierra
Caballo, El
Carmen, El
Central, La
Cocineras
Cristo, El
Dolores
Florencia
Inglaterra
Margarita
Mercedes
Mina Nueva
Mina Vieja
Parcionera
Peru
Potosí, El
Prieta
Quince (15) de Mayo
San Antonio
San Gregorio
San José
San Matias
San Toy
Santa Ana
Santa Gertrudis
Santa Juliana
Santa Rita
Santo Domingo
Sierra
Velardena

Minerals Reported

Actinolite
Alunogen
Andradite (?)
Anglesite
Aragonite
Argentite
Arsenopyrite, Auriferous
Azurite
Barite
Brochantite
Calcite
Cassiterite
Cerargyrite
Cerussite
Chalcocite
Chalcopyrite
Chlorite
Columbite
Copiapite
Cummingtonite
Descloizite
Dolomite
Embolite
Epidote
Epsomite
Fayalite
Fluorite
Galena
Gold, Native
Goslarite
Greenockite
Grossularite (?)
Gypsum
Gypsum, var. Selenite
Halite
Hedenbergite
Hematite
Hematite, var. Specularite
Hemimorphite
Hydrozincite
Idocrase (Vesuvianite)
Ilmenite
Ilvaite
Iodyrite
Kaolinite
Knebelite
"Limonite"
Magnesite
Magnetite
Malachite
Manganite
Marcasite
Melanterite
Mimetite
Mirabilite
Muscovite
Natrojarosite
Orthoclase
Proustite
Psilomelane
Pyrite
Pyrolusite
Pyromorphite
Pyrrhotite
Quartz
Quartz, var. Chalcedony
Quartz stained by Chrysocolla
Rhodochrosite
Rhodonite
Scapolite
Siderite
Silver, Native
Smithsonite
Sphalerite
Sphalerite, var. Marmatite
Sulfur, Native
Topaz
Tourmaline
Tremolite
Turgite
Uralite
Vanadinite
Wad
Willemite
Wolframite
Wulfenite

CUSIHUIRIACHIC DISTRICT

The Cusihuiriáchic District is an old silver camp in the Municipio of Cusihuiriáchic, Chihuahua. This District was discovered about 1666. Up to 1907, it produced more than 100 million dollars' worth of metals. This District is mined for gold, silver, copper, lead, and zinc. The Princesa Mine is located here.

Minerals Reported

Anglesite
Argentite
Azurite
Bornite
Bromyrite
Calcite
Cerussite
Chalcocite
Chalcopyrite
Galena
Hematite
Malachite
Polybasite
Proustite
Psilomelane
Pyrolusite
Quartz
Rhodochrosite
Sphalerite
Tetrahedrite
Wad

Barranca de Cobre

Copper Canyon

Copper Canyon, known in Spanish as *Barranca de Cobre,* is in western Chihuahua(s). This area is also known as the "majestic canyons of the Tarahumara," as it is in the Sierra de Tarahumara. Copper Canyon has been visited very seldom. It has deep canyons which are comparatively unexplored. The area around and in Copper Canyon has been mined for gold, silver, and copper.

WARNING: This area is completely isolated from civilization. Take plenty of gas, oil, water, and food. A guide is recommended.

The Pemex Travel Club publishes a good pamphlet on the *Barranca de Cobre*. It gives much interesting information on the canyon and its environs plus information on roads, facilities, and directions for reaching the area from Chihuahua or Los Mochis, Sinaloa. It is available free by writing the Club, or from tourist bureaus and consulates in the United States.

New Railroad

Mexican engineers, accomplishing an almost "impossible" project, completed a new railroad from Chihuahua to the Gulf of Mexico at Topolobampo in 1962. This traverses some of the most spectacular terrain of the Barranca de Cobre region and a trip on the new line is said to be a wonderful way to see the area.

BATOPILAS DISTRICT

The Batopilas District, Chihuahua, has been operated intermittently since 1632. A production of over 50 million ounces of silver has been recorded. The District, particularly the San Miguel Mine, has been noted for years as producing fine cabinet specimens of native silver. The native silver is found associated with calcite.

Batopilas District Mines

Animas	Nuestra Señora del Pilar
Arbitrios	Porfirio Díaz
Aurora	Sacramento
Ballinas	San Antonio
Camuchin	San Miguel
Carmen, El	Santo Domingo
Cinco de Mayo	Tajos
Cuatro de Julio	Tierra Blanca
Descubridora	Todos Santos
Gamuchino	Vavaca, La
Giral	Veta Grande
Mezquite	

Minerals Reported

Argentite	Pyrargyrite
Arsenic, Native	Pyrite
Arsenopyrite	Quartz
Calcite	Rammelsbergite
Galena	Safflorite
Polybasite	Silver, Native
Proustite	Sphalerite

NAICA

THE NAICA mining district is in the Sierra de Naica, Municipio of Saucillo, Chihuahua. The turn-off for Naica is approximately 68 miles north of Hidalgo del Parral, or 71 miles south of Chihuahua. Turn west from Highway 45 at kilometer 1545. There is a small railroad stop at the turn-off known as Estación Concho. It is about 15 miles farther by dirt road, almost impassable in wet weather. This district has been mined mainly for lead, zinc, and silver, with minor amounts of gold, copper, and mercury.

Like many mining districts in Mexico, Naica was at one time divided into a number of small mines operated by different companies. Now, all these smaller mines have been consolidated into a single unit operated by the Fresnillo Company. The former Fresnillo Company was nationalized in 1961, and is now known officially as Compañía Fresnillo, S.A.

Currently the mine is divided into two parts on the basis of different ownership: 1. the Naica mine, including the

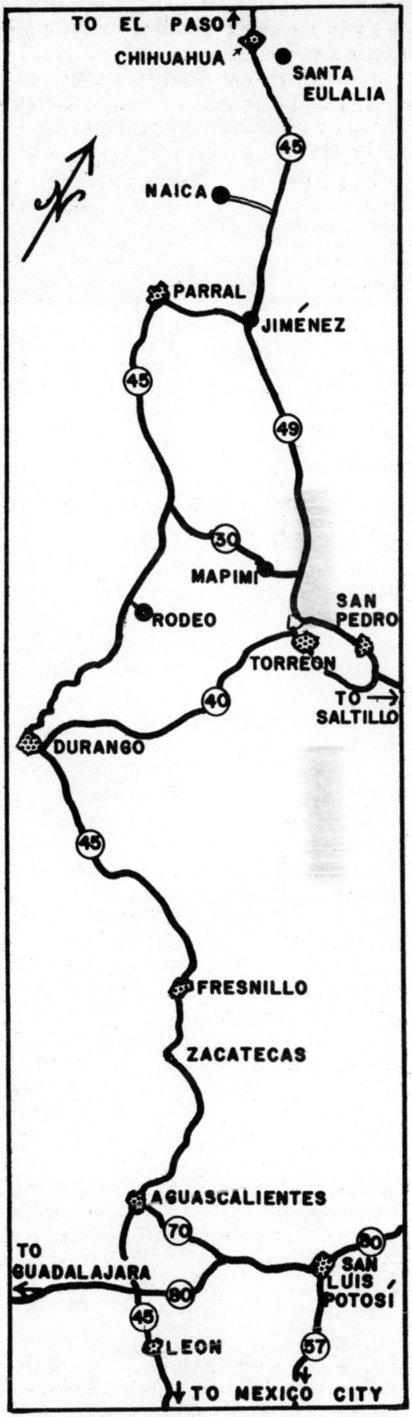

A fracture filled with selenite crystals in the Cave of the Swords.

former Maravillas, Ramón Corona, Lepanto mines, and several others, which since 1951 have been leased by Fresnillo from Minas de Naica, S.A.; 2. the Gibraltar mine which in 1956 was purchased outright by Fresnillo from the Eagle-Picher Company. The principal ore bodies in the Naica mine plunge into the Gibraltar mine at depth and most of the current development on the lower levels is in the Gibraltar mine.

The hospitable mine superintendents are H. F. Keller, general superintendent; and Juan Holguin, assistant general superintendent. Dr. Ivan F. Wilson is the resident chief geologist of the Fresnillo Company.

The Cave of the Swords

The Naica mine is world-famous as the locale of one of the truly magnificent natural wonders of Mexico — caves lined with selenite crystals. These caves were generally formed along the post-mineral fractures and, with one exception, to date have not been connected with the main ore bodies. The exception is a cave that was found inside the principal chimney ore body of the Gibraltar mine. This cave was lined with lustrous selenite crystals of the stubby type, some of which were three or four feet in length! Unfortunately this discovery had to be sacrificed to continue the mining operations. The largest of the caves is known as the Cave of the Swords.

To view the selenite caves, the visitor is first issued a hard hat, battery, belt, and miner's light. Before entering the mine, you are cautioned against touching the overhead electric wires, as your feet will be grounded on wet rail ties. This is to avoid electrocution. The light, casually strung over your right shoulder, casts a downward beam as you proceed through the Toledo tunnel, the main adit. The chances are that at least once a warning vibration will cause you to stretch your back against the tunnel sides while an ore train rumbles by in the darkness.

In time, the vertical shaft is reached, and you enter a small lift. The operator gives a signal tug on the cable and you start dropping at a rapid rate toward the interior of the mountain. It's "*aqui estamos,*" (here we are) at the 120 meter (390 feet) level.

The visitor, already subdued by the humid heat of a Turkish bath, sees and hears at a pumping station the torrent of water that registers 124 degrees Fahrenheit. This flood finds its source in the mine depths. Eventually it is pumped to the surface to serve the laundresses of Naica.

Continuing on, you slosh through the water that lies between the ore train tracks. Occasionally in the tunnel you pass through a large wooden door which is used to contain the choking dust from the blasting operations. If dynamiting is in progress, a wait of 45 minutes may be necessary. As each door is passed the temperature rises several degrees.

Very shortly you stop in the low, old tunnel. Finally, you come to a ladder and climb to a narrow entrance which is locked to prevent the miners from pilfering crystals. Once through the small door, your eyes open widely. There before you in all their pristine splendor,

The author (left), Nathalia Crane Black, and two miners in the humid heat, 500 feet below the surface at the Naica Mine, Chihuahua, Mexico.

are the long selenite crystals in the Cave of the Swords.

This cave defies description. The spectacular crystals of selenite, the crystalline variety of gypsum, grew from the sides, top and floor of this cave more than 500 feet below the surface. Walls with drawn swords — one false step and you would be impaled on a selenite dagger!

Some of the individual crystals in the rosettes which grew from the floor, reach a length of eight feet! The larger selenite

Selenite ,the crystallized variety of gypsum, from the Cave of Swords, Naica Mines, Chihuahua, Mexico.

crystals have a white cap on the termination and are light gray due to inclusions. In the upper part of the cave the selenite crystals have rounded terminations and have been partially redissolved, probably by natural causes. In the lower areas the crystals usually have well formed terminations. In 1950, the caves were flooded with hot water which may have had some effect on the crystals.

The Xochitl Cave is somewhat smaller than the Cave of the Swords and does not have large crystal clusters growing from its floor. Most of the Xochitl crystals are about 12 to 16 inches long. This cave has many exquisitely etched crystals with phantoms. Water inclusions with movable bubbles are common.

Both the Cave of the Swords and the Xochitl Cave are on the 390-foot level. Other caves have been found on some of the lower levels.

The caves are actually not too hot, only about 98 degrees Fahrenheit. It is the high humidity which wears you out. Perspiration pours from your skin. To prevent our cameras from being drenched with perspiration, we had to hold them in a raised position.

No one is allowed to remove or destroy the attractive crystals. Selenite crystals, however, can be purchased from Dr. Ivan F. Wilson at the Naica mine. The proceeds from these sales go into the miner's recreation fund.

Gypsum, Variety Selenite

Selenite, the transparent and generally colorless variety of gypsum, is found at the Naica mine in large crystals. These crystals occur in all types of ore but appear to be related to the period of oxidation. They were formed in cavities and open fractures just above the present water table.

The selenite crystals are of two basic types — long prismatic blades and short stubby ones. The stubby type rather than the long prismatic blades, have been found in all the lower levels of the mine below the 390-foot level. The unusual feature in the growth of selenite crystals at Naica is the great length in comparison with the narrow width. In a collection of prismatic selenite crystals, all under 15 inches long, the average ratio of maximum width to length on 34 measured specimens was 1:8. Very rarely a swallow-tail twin crystal is encountered, one on which the termination is V-shaped.

The selenite crystals probably were grown from supergene (descending water) solutions formed from the sulfuric acid given off by the oxidation of the sulfides. Surface waters also may have taken part in this reaction. The acidic solutions then reacted with the limestone to form calcium sulfate. Two water molecules were derived from the stagnant aqueous solutions and the selenite crystals started to grow. The growth process was evidently interrupted on several occasions and then continued again, as evinced by the many crystals with several phantoms in the same specimen.

The gypsum is sometimes covered by botryoidal or coral-like growth of calcite. This calcite may be drusy in habit and is often quite attractive.

Bassett (1955) has reported that gypsum crystals over a foot long were coated by calcite; later, solution of the gypsum core left unusual mold specimens. A similar type in which the gypsum crystals probably did not exceed two inches in length has been found. On these, white calcite in small drusy crystals first encrusted the gypsum. Then the gypsum was dissolved and finally the calcite was stained brown by iron oxides.

Visitors frequently express interest in curved selenite crystals. Any straight selenite crystal can be curved by heating slightly with your hands and then applying pressure until the desired shape has been attained.

Selenite Inclusions

Some of the large selenite crystals growing from the floor of the Cave of the Swords have black inclusions on the surface of phantoms. One such crystal grew first to three-fourths its present size, was encrusted by dirt, then started to grow again. At present the former surface can be recognized easily by a plane of inclusions which appear black at close range and give the crystal a gray cast from a distance.

Two-phase inclusions are commonly found in the crystals. They consist of liquid water of crystallization with a movable gas bubble. It has been rumored that in the past some of these crystals were used as natural levels.

A few of the crystals have light tan inclusions. One small cluster, evidently from one of the ore bodies, contains inclusions of pyrite. There are crystals which are brownish red due probably to inclusions of iron oxides. Small iron oxide inclusions at times are found near the crystal surface in selenite from the Cave of the Swords.

Some of the clear stubby selenite crystals from the Naica mine contain very interesting inclusions of cuprite. These small, bright red, cubic crystals are less than 1/16 inch across, are quite rare, and sometimes occur as groups of crystals growing next to each other. Some of these short crystals also contain native copper inclusions. So far as is known these inclusions are not found in the Cave of the Swords or Xochitl Caves.

One slightly curved, 14-inch-long selenite crystal from the Cave of the Swords, has a movable bubble which moves four inches, inclusions of iron oxide, six partial phantoms, and three full phantoms.

Rare Minerals

The rare minerals reported from the Naica mine include aurichalcite, bustamite, conichalcite, cosalite, matildite, and native mercury.

The mine, besides materially enriching the world's coffers in lead, zinc, and silver, will be known forever as the home of the Cave of the Swords — one of the most spectacular examples of crystal growth in the world.

Minerals Reported

Amphibole group
Andradite
Anglesite
Anhydrite
Argentite*
Arsenopyrite
Aurichalcite*
Azurite
Barite
Brochantite
Bustamite
Calcite
Carphosiderite*
Cerargyrite
Cerussite
Chalcocite
Chalcopyrite
Chlorite
Chrysocolla
Cinnabar
Conichalcite
Copiapite
Copper, Native
Cosalite
Covellite
Cuprite
Diopside
Dolomite
Fluorite
Fowlerite (?)
Galena
Galena, argentiferous
Gold, native
Grossularite
Gypsum
Gypsum, var. selenite
Halloysite
Hedenbergite
Hematite
Hematite, var. specularite
Hemimorphite
Hydrozincite
Idocrase (vesuvianite)
"Limonite"
Magnetite
Malachite
Manganhedenbergite
Matildite
Mercury, native
Mimetite
Molybdenite
Montmorillonite
Nontronite
Orthoclase, var. adularia
Plumbojarosite
Proustite*
Pyrargyrite*
Pyrite
Pyrolusite*
Pyrrhotite
Quartz
Quartz, var. chalcedony
Rhodochrosite
Scheelite
Silver, native
Smithsonite
Sphalerite
Sphalerite, var. marmatite
Tremolite
Wad*
Wollastonite
Wulfenite
Zoisite

*Mentioned in earlier reports.

Four-foot selenite crystals growing from the floor of the Cave of the Swords, Naica mine, Chihuahua, Mexico.

HIDALGO del PARRAL DISTRICT

The Hidalgo del Parral Mining District is near the City of Hidalgo del Parral in southern Chihuahua(s). The first silver veins were discovered in 1560. By 1632 the first mining had begun. The District is mined mainly for lead, zinc, and silver. There are two main mines. La Prieta Mine is owned by the American Smelting and Refining Company. The Esmeralda Mine is owned by a subsidiary of the Eagle-Picher Company.

The ore is in fault-filling veins in calcareous shale overlain by recent volcanics. The principal ore minerals are galena, sphalerite, chalcopyrite, and arsenopyrite. The gangue is composed of quartz, fluorite, and barite. Mineralization is localized along faults that cut the igenous rocks. The area also has a large monzonitic intrusion.

Santa Barbara District

The first fissure veins in the Hidalgo del Parral area were discovered at Santa Bárbara about 1547. The mines are grouped around the village of Santa Bárbara, about 15.5 miles by oil surface road southwest of Hidalgo del Parral.

This District is mined mainly for lead, zinc, and silver, with minor amounts of copper and gold. The Clarines Mine is owned by the San Francisco Mines of Mexico, Ltd. The Santa Bárbara Mine and all other mines in the area are operated by the American Smelting and Refining Company.

The ore of the Santa Bárbara District is found in fissure veins in calcareous shale which is slightly metamorphosed and intensesly folded. The shale is overlain by a volcanic cover which is more recent than the mineralization and includes rhyolite, diabase, and basalt.

The mineralogy of these mines can be divided into the sulfides and the oxides. The primary sulfides below the water table are mainly arsenopyrite, sphalerite, galena, chalcopyrite, and pyrite. The oxide zone above the water table contains anglesite, azurite, malachite, cerussite, jarosite, and "limonite." The principal gangue minerals are quartz, calcite, and fluorite.

San Francisco del Oro District

The San Francisco del Oro District is near the Hidalgo del Parral District, in the Municipio of Santa Bárbara, Chihuahua. It is 6.2 miles west of Santa Bárbara. This District was discovered about 1658. In terms of total silver production, the San Francisco del Oro District ranks third. It produces silver, lead, zinc, and a little copper and gold.

The ore is found in fissure veins in calcareous shale which is slightly metamorphosed and intensely folded. The shale is overlain by a volcanic cover which is more recent than the mineralization and includes rhyolite, basalt, and diabase.

There are several veins in the District, the most important of which is the San Francisco. The principal ore minerals include galena, sphalerite, chalcopyrite, and arsenopyrite. The gangue is comprised of calcite, fluorite, and high-temperature silicates.

Frisco Mine

The Frisco Mine in the San Francisco del Oro District, is the largest single mine in the Hidalgo del Parral area. It is 11.2 miles southwest of Hidalgo del Parral. The Mine is owned by San Francisco Mines of Mexico, Ltd. It produces daily 1,500 metric tons of ore, with a grade of 0.015 ounces of gold, 5 ounces of silver, 5% lead, 8% zinc and 0.6% copper per metric ton. The ore is found in fissure veins which occupy faults of small displacement in a calcareous shale. The sulfide minerals found in the veins include sphalerite, galena, chalcopyrite, pyrite, and arsenopyrite.

Hidalgo del Parral-San Francisco del Oro-Santa Barbara Mines:

Aguilarena, La
Alejandria
Almaloya
Alvaradena
Anexas
Apodaquena
Arembena
Balcequillo
Balleza
Bronces, Los
Cabadena
Caballo, El
Cabanitas
Cabos
Caiman
Caldas
Capitanena, La
Clarines
Cuadras
Cubana, La
Cuevas, Las
Embajadora, La
Esmeralda
Francia, La
Frisco
Huejotitan
Hueyuquilla
Jesús María
Labradena
Muertos, Los
Olivas
Palmilla, La
Perros Bravos
Plaza de Armas
Porvenir
Prieta, La
Quebradillas
Refugio
Ronces Valles
Rosario
Rota, La
Sabanera
San Antonio
San Cristóbal
San Diego de las Minas Nuevas
San Francisco del Oro
San José de Garcia
San Juanico
San Luis
San Patricio
San Pedro de la Cienega
San Rafael
Santa Ana
Santa Bárbara
Santo Tomas
Saynas
Sierra Madre
Tajo, El
Tecolotes
Terrenates
Tiro 8
Toro
Tuli
Unión, La
Valle del Parral

Minerals Reported

Actinolite
Anglesite
Argentite
Arsenopyrite
Azurite
Barite
Biotite
Bornite
Calcite
Cerargyrite
Cerussite
Chalcocite
Chalcopyrite
Chlorite
Chrysocolla
Copper, Native
Covellite
Datolite
Diopside
Diopside, var. Salite
Enstatite
Epidote
Fayalite
Fluorite
Galena
Garnet group
Gold, Native
Gypsum
Hedenbergite
Hematite
Hematite, var. Specularite
Hemimorphite
Herderite
Hisingerite
Idocrase (Vesuvianite)
Ilvaite
Jarosite
Kaolinite
"Limonite"
Magnetite
Malachite
Manganese oxides
Marcasite
Massicot
Mimetite
Muscovite, var. Sericite
Opal
Orthoclase
Orthoclase, var. Adularia
Plattnerite
Plumbojarosite
Proustite
Pyrite
Pyromorphite
Quartz
Quartz, var. Chalcedony
Scheelite
Silver, Native
Smithsonite
Sphalerite
Sphalerite, var. Marmatite
Stibnite (?)
Stilbite
Tetrahedrite
Tremolite
Vanadinite
Wulfenite
Zoisite

TALAMANTES DISTRICT

The Talamantes District is 12.4 miles east of Hidalgo del Parral, in the Municipio de Valle de Allende. It was discovered in 1890 and is the second largest producer of manganese in Mexico. The manganese occurs as veinlets, irregular nodular bodies, and irregular masses surrounded by rhyolitic breccia.

Talamantes District Mines

- Amalia 2
- Ampliacion de Amalia 2
- Ampliacion de la Reyna
- Estela
- Libertador, El
- Pavo Real, El
- Proteccion, La
- Reyna, La
- San Antonio
- Vida Nueva

Minerals Reported

- Barite
- Calcite
- Coronadite
- Cryptomelane
- Galena
- Gypsum
- Hematite
- Hollandite
- Psilomelane
- Quartz
- Quartz, var. Amethyst
- Quartz, var. Chalcedony

AMERICA-SAPIORIS (POTRILLOS) DISTRICT

The América-Sapiorís (Potrillos) District is in the Sierra de San Francisco, Durango(s). The Sierra de San Francisco lies along the northeast side of the plains of Guatimapé northeast of Patos, a village about 100 miles north-northwest of the City of Durango. This tin district is 9.3 miles long and 3.1 miles wide. The principal rocks are tin-bearing rhyolitic lavas, tuffs, and the breccias interbedded with them. The largest mines are the Grant and Candelaria. The very rare mineral durangite occurs here.

Mines near America

- Alta, La
- Barrocitas, Las
- Candelaria
- Carpas, Las (La Espesa)
- Fierrosa, La
- Grant
- Jaltomates
- Lentejuela, La
- Obsidiana mine of Herman West
- Orosco
- San Miguel
- Varocitos
- Veinte-nueve de Julio

Mines near Sapioris

- Gazapera
- Nueva (north of Sapiorís)
- Nueva (south of Sapiorís)

Minerals Reported

- Biotite
- Cassiterite
- Cristobalite
- Durangite
- Fluorite
- Hematite
- Hematite, var. Specularite
- Hornblende
- Magnetite
- Opal
- Orthoclase, var. Sanidine
- Pseudobrookite
- Quartz
- Quartz, var. Chalcedony
- Topaz
- Tridymite
- Zeolite group
- Zircon

MAPIMI

Mapimí in Durango(s), has long been known as the mineral collector's capital of Mexico. It is a small town with a population of about 3,120. The houses, constructed mainly of adobe, are clustered around one central plaza. It can be reached by a 58-mile drive east from La Zarca on Highway 45; or 38 miles northwest from Torreón on Highway 49.

Upon arrival in Mapimí it will take about 15 *seconds* for the word to be spread that *norteamericanos* have arrived. Then you will be deluged with friendly women and children selling *riscos* (specimens) from cardboard boxes. Everyone will want you to look at his selection. You will soon be completely surrounded. Patiently they await their turn to display their mineral wealth.

If you are collecting minerals professionally, a permit is required. The money received for the permit is used to help the hospital and poor people of Mapimí. The permit costs 10 to 20 pesos ($.80 to $1.60). When you pay for a mineral purchasing permit, be sure to get a receipt officially stamped. For those who intend to buy only a few minerals, a permit is not necessary.

Mapimí is world-famous as the home of La Ojuela Mine. Most of the minerals from that mine can be purchased in Mapimí, which may save you making the trip to the mine. The specimens purchased in town probably will be of much better quality than you can find on the dumps or in the mine. For further information about this area, see "Ojuela District."

OJUELA DISTRICT

The Ojuela District is near Mapimí in Durango(s). Its most famous mine is the Ojuela, one of the largest silver-lead deposits in Mexico. It produces lead, silver, and gold, with minor amounts of zinc, iron, and copper. The Ojuela Mine is in the Municipio of Mapimí, on the eastern side of La Bufa de Mapimí Mountain at the base of the northeastern escarpment of the Sierra de Mapimí which trends northwest-southeast. Bufa

Picture, page 28. Photomicrograph of spherules of velvet rosasite with aurichalcite and hemimorphite. Mapimi, Mexico.

(Cliff) Mountain rises more than 3,000 feet above the surrounding level plain. The name *ojuela* (little eye) is derived from a small hole resembling the eye of a needle, in the limestone near the top of Bufa Mountain.

Mining to shallow depths for rich secondary silver ores began here in 1598 and it has been continued intermittently to date. In the past, notched beams called "chicken ladders," were used to go from one level to another in the mine. In 1893, the Ojuela Mine was converted by Compañía Minera de Peñoles, S.A., a subsidiary of the American Metal Company, into an enormous lead-silver property. Since then more than five million tons of ore have been extracted.

The ore since 1893 has assayed on the average 3.7 grams of gold, 462 grams of silver, and 14.9 per cent of lead per ton. Large-scale operations ceased at the end of World War II in the face of mounting drainage difficulties and the depletion of reserves. In the past, 6,000 gallons of water per minute had to be pumped from the mine. Smaller operations are now carried forward by a mining co-operative using hand methods.

In the geologic past, the limestone and shale sediments of the area around Mapimí were domed by igneous intrusion. This was simultaneously accompanied by thrusting from the southwest. As a result of the thrusting and subsequent settling, the sediments were buckled into a series of overthrust anticlines and synclines. Faulting accompanied by a major system of fracturing and fissures controlled the mineralization.

The ore deposits occur in the limestone which is overlain by shale 100 feet thick. The limestone contains some dolomite and is called dolomitic limestone. It is a marine type, finely grained, and light gray. Some pure, thin-bedded dolomite beds above the limestone acted as a blanket or dam, confining the metallic solutions to the underlying beds.

The ore bodies at the Ojuela Mine are essentially chimneys or pipes with some mantos. With depth the ore bodies have a simple chimney structure. As they rise toward the surface they become more irregular. The mine has numerous individual chimneys.

To give some idea of a complex ore body in the Ojuela Mine, Prescott wrote, "San Diego starts as a strong manto below water level, crosses the post-mineral San Diego fault, mantos for one hundred meters more, goes up as a chimney for fifty meters, then follows the beds, crosses the San Diego fault, and continues on the beds, following faithfully the structure developed, a huge Z, and finally above the first level strikes the pre-mineral San Vicente fault, which it follows to the surface."

All of the underground tunnels on Bufa Mountain are known collectively as the Ojuela Mine. The whole mountain is underlain by tunnels. It is reported that there are 200 miles of underground workings. It certainly must be true that one could walk for weeks underground without visiting the same place twice.

The Ojuela Mine deposit, typical of those on the northern Mexican plateau, is a limestone replacement. Primary sulfide mineralization consisted chiefly of argentiferous galena, arsenopyrite, pyrite, and sphalerite. These are also the principal ore minerals. The small copper content of the ore increases with depth.

The principal ore bodies are widely separated by barren limestone and consist of enormous caves and pipes. Obviously all the ore body solutions worked upward from their source.

The matrix rock is composed of goethite, hematite, quartz, dolomite, and fluorite. The important gangue minerals

Vug filled with adamite crystals from La Ojuela Mine, Mapimi, Durango, Mexico.

The Americados (America number two) entrance to La Ojuela Mine, Mapimi, Durango. This was the only entrance being used in 1962.

with the ore are calcite, fluorite, goethite, and quartz. The Ojuela Mine is different from other silver mines in northern Mexico in that small pegmatite veins are found in its lower workings.

There are at least four main entrances to the Ojuela Mine, the Americauna (American number one), Americados (America number two), Norte (North), and Ojuela (Little Eye). Only one of the four entrances to the Ojuela Mine is being used, the Americados. To reach it, drive one mile east of Mapimí and turn to the right (south) toward the mountains. The entrance is three miles farther on. The dump at the Americados entrance contains arsenopyrite, calcite, chalcopyrite, galena, goethite, hemimorphite, malachite, scorodite, sphalerite var. marmatite, and wulfenite.

The amazing Ojuela entrance is about seven miles from Mapimí and three miles farther south than the Americados entrance. It is reached by the intrepid on a treacherous, one-way road carved out of solid rock. Four-wheel drive is a necessity!

The Ojuela entrance is the site of a deserted village where 2,000 people once lived. Atop the mountain, it supported a school, hospital, movie theater, dance hall, and many other conveniences. It had a cable car that one day plummeted 40 baseball players to their death in the valley below. Today all is in ruins. The only visible life left is small lizards basking in the sun.

The only structure still intact is one of the most spectacular man-made wonders in Mexico — an unbelievably long, cable suspension bridge. This bridge was constructed with the help of an American engineer. When we asked our guide whether the bridge was strong enough to support us, he replied, "If you reach the other side — the bridge was safe."

The rare minerals of the Ojuela Mine are adamite, austinite, bindheimite, chalcophanite, conichalcite, cuproadamite, duftite and rosasite. The very rare minerals include arseniosiderite, carminite, dussertite, legrandite, murdochite and paradamite. Because of the unusual mineral associations found at the Ojuela Mine, it may reveal several new mineral species in the future.

The Ojuela Mine will always be remembered by the connoisseur of unique mineral specimens. The unusual mineralogical associations found there are in themselves distinctive. Truly, the Ojuela Mine will go down in Mexican mining history as one of the most picturesque silver-lead-gold deposits.

Notes on Minerals Of the Ojuela Mine

You never fully realize how attractive are some of the Ojuela minerals until you observe them under the microscope.

Adamite

Can sometimes be found on the dumps at the Norte and Americauna entrances of the mine. Most of the adamite occurs as radial crusts and granular aggregations merged with small crystal faces. It is associated with calcite and chalcophanite. The matrix is goethite, hematite, or limestone.

Adamite commonly can be found crystallized in the vugs and cavities in the goethite and hematite. The colors include colorless, pale yellow, light yellowish green, and bluish green. Adamite from this area fluoresces yellow green under ultraviolet light. It shows a light yellow phosphorescence under short wave ultraviolet.

In the Las Palomas ore body, just above the eleventh level, a miniature grotto four feet in diameter completely lined with adamite crystals, was found about 1946 by two mineral collectors. Single adamite crystals in this large vug reached a length of 5/16 inch. A specimen of adamite three square feet, weighing 75 pounds, from the grotto can be seen in the U.S. National Museum. The author has two specimens of adamite in his collection from the Ojuela Mine with individual dipyramidal orthorhombic crystals over one-half inch long.

Adamite, var. cuproadamite

Occurs as light blue drusy crystals on the goethite matrix.

Anglesite

Is one of the less abundant ore minerals.

Aragonite

Has on rare occasions been found in some of the caverns in the upper workings.

Arseniosiderite

One of the very rare minerals encountered here is found finely grained massed with carminite. It is believed to be a pseudomorph after scorodite. At this locality it is a light to dark chestnut brown. It forms reticulating veinlets and replacement masses in scorodtie.

Arsenopyrite

Is a very common mineral disseminated through the ore.

Aurichalcite

One of the very delicate minerals found in the oxidation zone. These attractive light to dark blue specimens occur associated with rosasite, calcite, and hemimorphite on goethite matrix. It frequently occurs as acicular incrustations lining the vugs and cavities of the matrix. The aurichalcite is sometimes overlaid by a crust of hemimorphite or calcite crystals. It alters to the white mineral hydrozincite.

Austinite

A light green mineral associated with the darker green conichalcite. It is the zinc end member of the conichalcite-austinite series. Austinite is sometimes found in association with calcite on goethite or hematite matrix.

Barite

Can be found as white crystals associated with fluorite and wulfenite.

Bindheimite

Occurs as white to gray crusts on a goethite matrix. It is found in the oxidation zone associated with plattnerite.

Boulangerite

Reportedly is not abundant at the Ojuela Mine.

Calcite

Calcite with interesting inclusions of malachite can be found on the dump at the Americados entrance. Most of the calcite is white. It sometimes occurs as transparent modified crystals. Acute rhombohedral crystals, some in arborescent form, can be collected. Sometimes the calcite has inclusions of hemimorphite. It is associated with malachite on goethite or hematite matrix.

Carminite

Occurs in very small crystals less than 1/32 inch long, in small vugs or cavities in either scorodite or arseniosiderite. The crystals tend to group themselves in sheaf-like forms. It is associated with dussertite. Carminite has been collected on a small dump near the North entrance. It is highly lustrous and a dark carmine red.

Cerussite

In the past was one of the main ore minerals.

Chalcophanite

Occurs in blue black, drusy, botryoidal crusts and small tabular crystals on goethite. It is sometimes associated with adamite.

Chalcopyrite

Also is one of the less common ore minerals. It increases in abundance with depth.

Conichalcite

Is found as dark green botryoidal to reniform crusts and masses. It is the copper end member of the conichalcite-austinite solid-solution series. Conichalcite is found in small crystals incrusting the stalactitic goethite or hematite. Under the dark green conichalcite is sometimes found the light green austinite. Conichalcite is usually associated with calcite.

Dolomite

Occurs as almost pure veins and is associated with the Aurora limestone.

Duftite

A rare mineral, has been discovered coating wulfenite crystals. It is olive-to gray green.

Dussertite

Can be found in small flattened crystals sometimes grouped as rosettes or crusts. It occurs in scorodite associated with either carminite or arseniosiderite. Dussertite is light green and has a fine granular texture.

Fluorite

Appears as cubic crystals and octahedral cleavage fragments. The color varies from clear to wine red and purple.

Galena

Sometimes argentiferous (containing silver), is present as an abundant ore mineral.

Goethite

Is a common matrix at the Ojuela Mine. It was formerly called "limonite."

Hematite

Forms part of the matrix and is associated wtih goethite and "limonite."

Hemimorphite

Occurs as white tabular crystals which often form sheaflike aggregates. It is sometimes found alone crystallized on the goethite matrix or in association with aurichalcite, rosasite, or calcite. Hemimorphite is occasionally encountered with interesting inclusions of aurichalcite. It often incrusts other minerals, notably aurichalcite, or cigar-shaped goethite.

Wulfenite from La Ojuela Mine, Mapimi, Durango.

Evidently it was one of the last minerals to form in the oxidized zone.

Hydrozincite

Is found as delicate white acicular crystals on goethite matrix. It is formed as an alteration product of and associated with aurichalcite.

Jarosite

Is an amber-yellow to dark brown, secondary mineral usually found as crusts or coatings on ferruginous ore.

Legrandite

Occurs as radiating aggregates of prismatic crystals up to about two inches in size; also as small, radiating, fibrous to acicular crystals which are usually less than 1/16 inch. Legrandite is associated with adamite, calcite, jarosite, pyrite, quartz, and smithsonite. It is usually found on hard goethite ("limonite") matrix. It is colorless to bright canary-yellow. An extremely rare mineral.

"Limonite"

The name has been used for years to describe the matrix at the Ojuela Mine. This mineral has generally been shown to be goethite. The name "limonite" still serves a purpose in the field to describe earthy goethite.

Malachite

The green copper carbonate occurs

as acicular crystals on goethite. It is associated with calcite and plattnerite, and is sometimes overlain by white calcite.

Mimetite

Can be found as small white, pale yellow, or yellowish brown globular stalactitic crystals. They occur on goethite matrix associated with wulfenite and white calcite. It forms a series with pyromorphite.

Minium

This bright red mineral is of secondary origin, derived from the alteration of galena or cerussite. It is earthy in form.

Murdochite

Occurs as black opaque octahedral crystals less than 1/32 inch in size. It is associated with calcite and plattnerite and can sometimes be found by searching the latter with a microscope.

Paradamite

Is a new triclinic zinc arsenate first found at the Ojuela Mine. It is a dimorphic form of the orthorhombic adamite. Paradamite is also a paramorph (different internal molecular structure, with the same chemical composition and external form) of adamite. It occurs as sheaflike aggregates of crystals and as somewhat rounded and striated equidimensional crystals up to five millimeters (3/16 inch) in size. They are transparent, have a vitreous luster, and are pale yellow. Paradamite is associated with mimetite and adamite on goethite matrix. It has a perfect cleavage, a specific gravity of 4.55, and a white streak.

Plattnerite

Frequently is encountered at the Ojuela mine. It is black and is associated with calcite, murdochite, hemimorphite and malachite. At this locality it occurs as sharp shiny crystals of microsize.

Plumbojarosite

Can be found with the ore as light to dark brown compact masses.

Pyrargyrite

Called ruby silver, is very rare at the Ojuela Mine.

Pyrite

One of the common ore minerals.

Pyrolusite

This black manganese oxide is not encountered frequently at the Ojuela Mine.

Pyromorphite

Is green or yellow and forms a series with mimetite.

Quartz

Sometimes is found in small amounts in the ore.

Rosasite

Occurs in beautiful, blue, velvet, mammillary spherules on goethite matrix. It is sometimes associated with aurichalcite and hemimorphite. Occasionally the single rosasite spherules are incrusted by hemimorphite crystals.

Sauconite

A white clay, possibly has been discovered associated with aurichalcite and hydrozincite.

Scorodite

Forms large masses and occurs in coarse, granular, and rounded concretionary forms. It is also found on arsenopyrite as a light yellowish green alteration product of that mineral. Scorodite is associated with arsenopyrite, marmatite, arseniosiderite, carminite, and dussertite.

Native Silver

Is reported as very rare at the Ojuela Mine.

Smithsonite

Is found sporadically with the principal ore minerals.

Sphalerite

Can be collected in white-yellow-gray, massive botryoidal form. Sphalerite, var. marmatite, contains from 10 to 30 per cent iron. It is one of the common ore minerals.

Stibiconite

Is the main mineral in the antimony ocher. It is not abundant at this locale. It is pale yellow to yellowish white and occurs as a powder, compact, or massive.

Stibnite

Very rarely is found at this mine.

Tetrahedrite

Is an uncommon mineral at this locality.

Wulfenite

In sharp contrast to the thick, square, Los Lamentos crystals, wulfenite occurs at the Ojuela Mine as very thin, square, tabular, light yellow crystals. It is associated with calcite, mimetite and plattnerite. The width of each delicate crystal is usually less than ⅝ inch.

Profile of the Cerro de Mercado mine, Durango, Durango. Also known as "iron mountain."

VELARDENA DISTRICT

The Velardeña District is between Torreón and Durango, Durango, on Highway 40. The District is mined for silver, gold, and lead. The new species, hillebrandite, krausite and spurrite, were first found here.

Velardena District Mines

Guardarraya	Terneras
Santa María	

Minerals Reported

Actinolite	Magnetite
Apatite	Merwinite
Arsenopyrite	Mimetite
Augite	Orthoclase
Barite	Orthoclase, var. Adularia
Biotite	Pyrite
Calcite	Pyrite, Cupriferous
Chalcopyrite	Pyrrhotite
Chlorite	Quartz
Diopside	Quartz, var. Chalcedony
Epidote	Rutile
Fluorite	Serpentine
Galena	Sphalerite
Galena, Argentiferous	Spurrite
Garnet group	Tetrahedrite
Gehlenite	Tetrahedrite, Argentiferous
Halotrichite	Tetrahedrite, Auriferous
Hedenbergite	Titanite
Hematite	Tourmaline
Hematite, var. Specularite	Voltaite
Hillebrandite	Wollastonite
Hornblende	Zircon
Idocrase (Vesuvianite)	
Jamesonite	
Krausite	

DURANGO

The City of Durango (doo-RANG-go) is still steeped in colonial tradition. Although it has a population of over 95,000 and is the capital of the state of the same name, it still retains the charm of old Mexico. Of interest to the mineral and gem collector is the large iron deposit at the edge of the city.

In 1552, the Governor of New Galicia (Jalisco) sent Ginés Vásquez de Mercado in search of gold, silver, and new lands. Upon nearing the site of the present City of Durango, Sr. Mercado noticed a hill rising 700 feet above a level plain which extended for miles. He excitedly climbed the hill only to find it was almost solid iron instead of silver or gold. It remained untouched until 1828 when small operations commenced. Large-scale production began in 1888.

The Cerro de Mercado (Mercado's Hill) Mine is sometimes called "Iron Mountain." A hill of solid iron ore, it is one of the largest single iron deposits in existence. The Cerro de Mercado Mine is Mexico's chief domestic supplier of high-grade iron ore. It is a typical replacement deposit. The silica-rich extrusive rocks are of three types: latite, rhyolite, and rhyolitic tuff.

The principal ore mineral is martite,

Apatite crystals from Durango, when clear and without flaws or fractures, can be faceted into attractive gemstones.

a variety of hematite, pseudomorphous after magnetite. It occurs in large octahedrons. Goethite is also a dominant mineral.

The ore which assays 60 to 62 per cent iron, is drilled and blasted. It is then skip-loaded into dump trucks. Finally it is crushed and shipped by railroad to the smelter. Over 300 tons of ore are mined per day and at this rate the reserves are estimated to last for another hundred years.

Attractive yellow apatite crystals long have been noted from this locality. They are called *amarillas* (yellows) by the miners. These crystals are usually less than one and three quarters inches wide and two inches long. They sometimes have a brown iron oxide stain on the surface. It is customary to clean this stain from quartz with oxalic acid. *This will not work with the apatite,* as it is a phosphate and will be attacked by the action of the acid. Unfortunately, most of the yellow apatite crystals are fractured in the process of blasting the ore. It is possible to buy high-graded apatite from the miners in Durango. A clear crystal can be faceted into a colorful gem.

Engineer Macario Muñoz is the local director of the Comisión de Fomento Minero. His office is at Victoria Sur No. 109, Durango. He has a representative mineral collection in his office from the State of Durango, and Mexico in general.

In Durango, optical calcite can be purchased from Adolfo Torrecillas who lives at Baca Ortiz No. 108. It is quite expensive and one specimen costs about $7.00. Optical calcite has many uses in optics, bomb sights, etc. This calcite is from his El Amparo Mine, Rodeo, Durango.

Although no gold or silver was found at Iron Mountain, it has proved extremely important to the Republic of Mexico.

Minerals Reported

Apatite	Magnetite
Augite	Mountain leather (Amphibole asbestos)
Barite	Opal, var. Hyalite
Calcite	Pyrite
Dahllite (Carbonate-Apatite)	Quartz
Goethite	Quartz, var. Chalcedony
Hedenbergite	Sepiolite
Hematite	Titanite
Hematite, var. Martite	
Hematite, var. Specularite	

CERRO de los REMEDIOS DISTRICT

The Cerro de los Remedios District is a tin-mining area at the southwest edge of the City of Durango.

Minerals Reported

Biotite	Opal
Cassiterite	Orthoclase, var. Sanidine
Cristobalite	Pseudobrookite
Fluorite	Quartz
Hematite	Quartz, var. Chalcedony
Hematite, var. Specularite	Topaz
Magnetite	Tridymite
Montmorillonite	Zircon
Oligoclase	

SOMBRERETE DISTRICT

The Sombrerete District is some 6.2 miles north of Sombrerete, Zacatecas. It was discovered in 1546. The country rock at Sombrerete consists of marly limestone of Mesozoic age, shale, and graywacke. Some of the hills are covered by Tertiary rhyolite. The ore is found in veins. The metals produced, in order of abundance, are silver, gold, lead, and copper.

Minerals Reported

Argentite	Pyrite
Calcite	Quartz
Chalcopyrite	Scheelite
Galena	Silver, Native
Miargyrite	Sphalerite, var. Marmatite
Polybasite	Stephanite
Priceite	
Pyrargyrite	

FRESNILLO DISTRICT

Fresnillo, Zacatecas, lends its name to the mining district, the mine, and the company that operates the mine. The Fresnillo Mine owned by the Fresnillo Company, is primarily a silver, lead, and zinc producer with small amounts of copper and gold. It was first opened in 1554. The District is first in Mexico in silver production. The mine employs 2,800 men and has been exploited to a depth of 3,500 feet.

The Fresnillo Company also leases the Naica Mine at Naica, Chihuahua. The general manager of the company is Fred W. Bailey.

The deposits [illegible] occur in two forms, fissure veins and stockworks. The ore veins occur in normal fault fillings and veins which have been reopened and filled. The normal veins average less than four feet wide. One unusual vein is called the *Cueva Santa* vein and is 15 to 18 feet wide. The ore averages about 200 grams of silver per ton.

The main minerals in the ore include chalcopyrite, galena, sphalerite, var. marmatite, and pyrite, with minor amounts of pyrrhotite, arsenopyrite, and pyrargyrite. The silver-bearing minerals found in the ore are pyrargyrite, proustite, and argentite. The gangue is composed mainly of quartz and calcite, with minor amounts of barite. The one unusual mineral found here is axinite.

Galena and chalcopyrite from the Fresnillo Mine, Fresnillo, Zacatecas, Mexcio.

The Fresnillo Mine has trouble with water. They are trying to drain one tunnel by pumping. At the bottom of the mine the temperature is 102 degrees Fahrenheit, the humidity is about 98 per cent.

The Mine comprises mainly three lithologic units: 1, upper graywacke, consisting of argillite and thin-bedded limestone; 2, calcareous shales; and 3, lower graywacke consisting of argillite only. There are also basalt and rhyolite flows, interbedded tuffs, sediments, etc.

On the tailings dump are 30,000,000 tons of material. The oxides are on the bottom and the sulfides on top. There are no specimens on the dump; however, they can be obtained from the ore cars at the mill.

The mill superintendent at the Fresnillo Mine is Charles J. Veale. Robert L. Meyer is one of the geologists for the company, and Dr. Ivan F. Wilson at Naica, Chihuahua, is the chief geologist.

The operation of the mill at Fresnillo is quite interesting. First the ore is crushed in a coarse crusher from thirty inches down to four. In a secondary crusher it is reduced to one and one-half inches. Then it is crushed in the tertiary crusher to less than five-eighths of an inch. Next, the five-eighths-inch ore is crushed by iron bars. Finally, in the ball mill (which uses iron balls) the ore is crushed to a fine powder. The flotation process is then used to recover the concentrates.

The flotation method was introduced at Fresnillo in 1926, whereby they use cresylic and xanthic acids, zinc sulphate, and sodium cyanide. The chemicals and ore are stirred up by air which raises bubbles. The densest material adheres to the bubbles by surface tension and is taken to the surface where a froth is formed. The froth is collected and the concentrates are recovered.

From every hundred tons of ore, four tons of lead concentrate are recovered. The mine produces 1,400 to 1,500 tons of concentrates per day, which are shipped to the smelter to be reduced further.

The Fresnillo Mine has produced more

These mountains around Guanajato have yielded many fine mineral specimens.

than 200 million ounces of silver, chiefly from oxidized ores. Total production at the Fresnillo Mine from 1554 to 1956 is shown below.

Gold	26,455 pounds
Silver	17,085,650 pounds
Lead	450,000 metric tons
Zinc	400,000 metric tons
Copper	45,000 metric tons

At current prices this production would be worth more than 260 million dollars.

Minerals Reported

Argentite	Polybasite
Arsenopyrite	Proustite
Axinite	Pyrargyrite
Calcite	Pyrite
Chalcopyrite	Pyrrhotite
Diopside	Quartz
Galena	Sphalerite
Hedenbergite	Sphalerite, var. Marmatite
Matildite	

ZACATECAS DISTRICT

Zacatecas, Zacatecas, (zah-kah-TAY-kahs) is one of the world's most famous silver camps. It has been mined since 1548 for lead, zinc, silver, copper, and gold. It is also famed for its beautiful colonial aqueduct. The hills around Zacatecas are honeycombed with mines both abandoned and in operation. Collecting may be done on any of the abandoned mine dumps.

In Zacatecas, the district office of the Comisión de Fomento Minero is located at Ignacio Hierro No. 4. Engineer Aurelio Martinez is in charge of the office here and in San Luis Potosí and alternates between the two cities.

The rocks of the Zacatecas District include schist, quartzite, conglomerate, limestone, diorite, diabase, andesite and rhyolite. In the oldest rocks occur important anticlinal and synclinal folds. Variously shaped ore bodies have been found. The most important ore body structures are veins, faults, and broken zones. Most of the ore is found in fissure veins.

Geologists believe there were three periods of primary mineralization, first, veins of copper and gold; second, veins of silver, lead, zinc, and gold; and finally, veins of gold and sliver.

Sphalerite is found in the Zacatecas District with inclusions of chalcopyrite. Quartz crystals are found with inclusions of sphalerite.

Zacatecas District Mines

Abundancia, La	Parroquina
Anaconda	Quebradilla
Bilbao	Refugio
Bote, El	San Bernabe
Cantera	San Jesús
Carmen	San Rafael
Guadalupe	San Roberto
Magistral	Zaragoza
Mala Noche	

Minerals Reported

Anglesite	Hematite
Argentite	Kaolinite
Azurite	"Limonite"
Barite	Malachite
Bornite	Miargyrite
Bournonite	Mimetite
Bromyrite	Muscovite, var. Sericite
Calcite	
Cerargyrite	Polybasite
Cerussite	Psilomelane
Chalcocite	Pyrargyrite
Chalcopyrite	Pyrite
Copper, Native	Pyrolusite
Covellite	Quartz
Cryptomelane	Quartz, var. Chalcedony
Descloizite, var. Cuprodescloizite	Rhodochrosite
Dolomite	Silver, Native
Fluorite	Sphalerite
Freieslebenite	Stephanite
Galena	Tetrahedrite
Galena, Argentiferous	Tetrahedrite, var. Freibergite
Gold, Native	
Gypsum	

TEPEZALA DISTRICT

The Tepezala District is 10.6 miles east of Rincón de Romos, Aguascalientes. It was first mined by the Spaniards. The District is mined for copper, lead, zinc, and silver. The ore is found in veins which cut limestone and rhyolite. The principal ore minerals are chalcocite, chalcopyrite, galena, and sphalerite. The gangue associated with the sulfide ore is composed of diopside, hedenbergite, epidote, and quartz.

Minerals Reported

Actinolite	Grossularite
Aragonite	Hedenbergite
Calcite	Hematite
Chalcocite	Magnetite
Chalcopyrite	Malachite
Chlorite	Pyrite
Diopside	Quartz
Epidote	Sphalerite
Fluorite	Uralite
Galena	Zoisite

GUANAJUATO DISTRICT

Nestled in the hills of the central Mexican plateau is the City of Guanajuato (gwah-nah-HWAH-toh), capital of the State of Guanajuato. It was founded in 1554. It even has some cobblestone streets as a reminder of its colorful heritage. The Guanajuato District has been mined since 1548.

The early mining was done by building a fire along the face to be advanced and then quenching it with cold water. Later, mines employed the Buscon system. A company owned the mine and opened the shafts, adits, drifts, etc. All or part of the ore extraction was turned over to a lessee or *buscon*. The buscon hired his own men and extracted the ore at his own expense. The only thing he was provided was a place to work. The ore was divided into two equal piles, one for the company, the other for the buscon.

The mineral deposits occur in shale, metamorphic rock, red conglomerate, and even extrusive rocks (andesite and rhyolite). Limestone and tuffs are also found in the District. The ore is of two main types, well defined veins which occupy fault planes, and stockworks. There are three vein systems in the Guanajuato District: Sierra,Veta Madre, and La Luz.

A great octagonal shaft called El Tiro General, was started in 1801. Shafts were built with eight sides to facilitate loading from eight different adits (tunnels) at the bottom. Many of the mines are filled with water in the lower levels which extend below the water table. The water has to be pumped out at great expense or the operations discontinued.

The Veta Madre vein system was discovered in 1550. Geologists have claimed that it has been the richest gold-silver vein in the entire world. Even today this vein is worked on a reduced scale. Before 1927, the vein was estimated to have produced more than one billion dollars' worth of metals.

The Veta Madre occupies a large fault zone, and, along with conglomerate, is in contact with shale, schist, rhyolite, and andesite. This lode, the most important in the Guanajuato District, is about 16 miles long and varies in width from 12 to 130 feet. The ore is concentrated in shoots 600 to 1,200 feet long. Most of the ore bodies are contained in a three-mile length along the vein.

Guanajato is nestled in a canyon.

The remains of the famous La Valenciana Mine is on the hill above Guanajuato. All that are left of what was one of the greatest silver producing mines in history are the dump and a deep, eight-sided, vertical shaft. It had its bonanza from 1768 to 1810.

The mine entrance is surrounded by a thick wall. A family lives inside the ruined enclosure and sometimes has specimens for sale. The high walls and expanse of the courtyard are typical of Spanish architecture. Apparently the ore was smelted at the mine, for there are ruins of smokestacks still visible in the rubble.

La Valenciana shaft drops straight down 1,650 feet. Most of the silver mines in Mexico have a vertical shaft through which the ore is hoisted from horizontal adits at the bottom. This mine was closed about thirty years ago.

An operating silver mine can be found near La Valenciana. The Tiro Tepeyac mine is owned by the Sociedad Cooperativa Minero-Metalurgica, Santa Fe de Guanajuato No. 1, Guanajuato. Sr. Pedro Rodriguez is in charge. The ore from this mine which is over 600 feet deep, assays approximately two grams of gold and 350 grams of silver per ton.

The University of Guanajuato has a fine collection of minerals, especially those found locally. The collection is badly in need of new labeling.

The big tourist attraction in Guanajuato is the Municipal Pantheon or burial ground where the mummies are kept. Depending upon the viewer's temperament, the mummies are either interesting or terrifying. Two of the caretakers at the Pantheon have mineral specimens for sale. They offer mostly amethyst. A 5x6-

inch specimen usually sells for about 10 pesos (80 cents). Most of the amethyst is from the San Juan de Rayas mine in Guanajuato.

There is one mineral dealer, Sr. Garcias, in Guanajuato. He sells mostly quartz, amethyst, and a small amount of calcite. Anyone in town can direct you to the man who sells *piedras, muestras,* or *riscos* (riches), as minerals and gems are called in Spanish.

Three minerals new to science were discovered in the Guanajuato District — aguilarite, guanajuatite and paraguanajuatite. The rare minerals of the Guanajuato District include those three and naumannite.

Guanajuatite has been found at the Santa Catarina and La Industrial mines, and at Sierra de Santa Rosa. Santa Rosa is a few miles north of Guanajuato, on the road to Dolores Hidalgo. Paraguanajuatite is reported from the Santa Catarina mine.

Very rarely, skeletal, dodecahedral crystals of aguilarite have been found associated with quartz, imbedded in colorless calcite in the San Carlos mine.

Apophyllite in beautiful pink and white crystals has been found in the El Refugio mine.

In the late nineteenth century, Dr. A. E. Foote bought crystals of calcite in Guanajuato. Apparently all of these crystals came from the same cavity. When viewed from the side, these unusual crystals appear to have a white line running lengthwise through the center. The "white line" is a small rhomboidal tube through which a hair can be passed. It is parallel to the sides of the crystal. Close to the crystal termination the tube narrows and apparently disappears. The tubes could be considered negative crystals.

If you go to Mexico, be sure to include Guanajuato in your itinerary. You will be sure to find it a rich and rewarding experience historically, scientifically, and mineralogically.

The headframe at the Tiro Tepeyac mine, Guanajato, Mexico.

Guanajuato District Mines

Bolanitos
Cata, La
Industrial, La
Jesús María
Locos, Los
Luz, La
Melladito
Mellado
Nueva Luz
Purisima, La
Refugio, El
San Bernabe
San Carlos
San José de los Muchachos
San Juan de Rayas
San Pedro
San Vicente
Sangre de Cristo
Santa Anita
Santa Catarina
Santa Clara
Santo Nino
Secho
Sirena
St. Nicholas
Tiro Tepeyac
Trinidad, La
Valencia
Valenciana, La

Minerals Reported

Aguilarite
Ankerite
Apophyllite
Argentite
Arsenopyrite
Barite
Bismuth, Native
Bismuthinite
Bismutite
Calcite
Celestite
Cerargyrite
Chabazite
Chalcopyrite
Chlorite
Cinnabar
Datolite
Dolomite
Embolite
Epsomite
Fluorite
Galena
Gold, Native
Guanajuatite
Gypsum
Gypsum, var. Selenite
Halotrichite
Heulandite
Laumontite
"Limonite"
Marcasite
Miargyrite
Molybdenite
Naumannite
Opal, Precious
Orthoclase, var. Adularia (Valencianite or Adularite)
Orthoclase, var. Sanidine
Paraguanajuatite
Polyargyrite
Polybasite
Proustite
Psilomelane
Pyrargyrite
Pyrite
Pyrrhotite
Quartz
Quartz, var. Amethyst
Quartz, var. Chalcedony
Raumonite
Rhodochrosite
Saponite
Siderite
Silver, Native
Sphalerite
Stephanite
Stibnite
Stilbite
Tennantite
Tetrahedrite
Topaz

El Santin Mine

El Santin mine is near Santa Catarina, a small village in the valley of the Tierra Blanca River. The village is 42 miles by road northeast of San Luis de la Paz, Guanajuato. The mine is about five miles south-southeast of Santa Catarina. There is no road to the mine.

The geology of the area consists of Cretaceous limestone overlain by at least 2,000 feet of volcanic rocks. Cassiterite is found in veins and placers derived from the veins. The mine is on the south side of the Mesa de la Fajas in the southern part of the Sierra Gorda.

Minerals Reported

- Apatite
- Biotite
- Cassiterite
- Cristobalite
- Fluorite
- Hematite
- Hematite, var. Specularite
- Magnetite
- Montmorillonite
- Opal, var. Hyalite
- Ordoñezite
- Orthoclase, var. Sanidine
- Plagioclase
- Pyroxene group
- Quartz
- Quartz, var. Chalcedony
- Topaz
- Tridymite
- Zeolite group
- Zircon

San Miguel de Allende

While in the vicinity of Guanajuato or Querétaro, a visit should be paid to San Miguel de Allende. This is one of North America's most famous literary and art centers. It is also an historic area where many of the events in the history of Mexican independence took place. Phil Kendall, a *norteamericano*, has a small gem and mineral shop a few doors from the plaza. He handles opal and other Mexican materials.

A good route from Guanajuato to Querétaro is via San Miguel de Allende and Dolores Hidalgo.

Bargaining for a jar of rough matrix opal at Trinidad, near San Juan del Rio, Queretaro.

QUERETARO

Querétaro, Querétaro, (kay-RAY-tah-ro) is in the heart of the Mexican opal district. There are several mineral and gem stores in town and several lapidary shops where opals are cut and sold. Also, there is no lack of street vendors who offer everything from genuine opal to "genuine alexandrite from Alexandria," including faceted pop bottle bottoms. The best course is to buy only from established dealers.

There are many home lapidaries in Querétaro, operated by one man or a family. There are a few organized shops that supply stones for the trade though they will also sell them directly.

Good opal usually can be purchased from Eugenio Ontiveros, 15 de Mayo 86, Querétaro. Sr. Ontiveros is well established and offers most of the gem materials to be found in Mexico, as well as American and Swiss synthetics. Most of them are cut in his own shop. Much of the opal is from La Carbonnera opal mine near Trinidad, which he owns and operates. He also has rough material and usually some mineral specimens.

Arnulfo Garcia, Madero 5, Querètaro. is another dealer. The extent of his stock is not known.

Good gems can be purchased from Señor Abbot. His shop is in his home on the highway which runs along the south edge of the city.

Mexican Opal

Opal, with the exception of agate, is the only gem mineral that is exploited commercially in any quantity in Mexico. Opal was discovered about 1855, reputedly at the Santa María Iris mine on the Hacienda Esmeralda about 15 miles east of Querétaro.

Opal is mined in the States of Querétaro, Hidalgo, Guerrero, Michoacan, Jalisco, San Luis Potosí, Chihuahua, and Guanajuato. It is generally conceded that the best Mexican opal comes from Querétaro(s) though the area around Magdalena, Jalisco, is now producing some gems of equal quality. (See Magdalena, Sec. 1.)

Querétaro opal is usually orange and yellow, rarely blue, with flashess of red and green. Most of it has the characteristic limpidity of Mexican opal. Some of it is completely transparent and can be faceted.

Opal usually is mined from open quarries. The matrix is generally hard rhyolite. Because of the hardness of the matrix, blasting is necessary and many fine stones have been ruined by the blasts. Sometimes the opal completely fills cavities in the rhyolite. At other times it is nodular and found loose in vugs and cavities, though rarely.

Watch out for opal doublets. A great many doublets are made with obsidian backing. They are easy to detect if unmounted, but beware of mounted stones where the edge or back cannot be examined. Many of the doublets are beautiful. The limpid opals against the black obsidian gives some of them the appearance of black opal. There is nothing wrong with opal doublets as such, but they should be lower in price than solid opals of the same size.

The highest quality Mexican opal is expensive. Very little good material can be obtained at the mines. Fine opal can be purchased in Querétaro, Mexico City, or Taxco, but the price will be as high as that of any comparable quality opal, regardless of origin.

Reputable dealers sell some "guaranteed" or "aged" opal. This is opal that has been mined, cut, and held for several years. The stone has been stored and allowed to dehydrate thoroughly to see whether it will check, craze, or lose its fire. Only the highest quality opal receives this treatment. When opal is offered with a guarantee by a reputable dealer, you can be sure it will be as stable as any opal.

Most rough opal and some cut stones are stored in water, glycerin, or mineral oil. In the case of rough material, this is to show it to best advantage. In cut stones, it is usually because the owner fears or knows they will dehydrate, crack, or lose their fire. Beware of cut stones in liquid. Buy them at your own risk. If you do, they should be quite cheap, as it will be you who is taking the chance.

No fixed price can be set on opal. It depends on the quality and the ability of the buyer or seller to bargain. Prices of cut stones will range from as low as 50 cents per carat to $50, $100, or more per carat. Fine stones are priced as high as several thousand dollars. These prices are for high quality, superb opal that will rival any in the world.

In most markets, particularly at the mines, you will find lower grade opal and opal in matrix. The rhyolite matrix is practically as hard as the opal. When the opal is small or irregular in shape, it is frequently left in the matrix; to try to remove it would risk destroying the opal. This opal is cut as cabochons and more or less of the matrix is left in place around the opal. The matrix cuts and polishes well and sometimes these stones are very attractive. Depending upon the quantity of opal that shows in relation to the matrix, such stones can be purchased for as little as five pesos (40 cents) to 20 pesos or more.

Rough opal is offered both as good, clear, cuttable pieces and in matrix. The clear pieces will vary in price according to quality and shape. Many rough Mexican opals have very odd shapes and cannot be cut economically. When buying, be sure to visualize how the piece will be cut and how much will be lost in the process.

Matrix opal purchased at or near the mines is fascinating material. It is sometimes sold by the piece but usually in jars varying from quarter-pint to quart size. The sellers are not ignorant. They not only know good opal and its value, but are shrewd enough to pack the jars with the good pieces on the outside. It's a case of *caveat emptor* — let the buyer beware! Jars can be bought for from as little as one peso to 10, 20, or more pesos. You make your own deal.

Opal Names

Abanderos — flame-colored precious opal. *Bandera* = flag, banner.

Arananjados — a name for amber-colored fire opal.

Azules — transparent with blue opalescence and vivid flecks of burning red and cooling green.

Cherry Opal — a red variety of common opal.

Common Opal — one without a play of color.

All you have to do to find precious opal is crack a few tons of this hard, rhyolitic rock at La Carbonera mine.

Contra Luz — precious opal which shows the best color when viewed by transmitted light.

Fire Opal — red body with red and green fire. The base color ranges from wine-yellow to yellow, cherry, and brown.

Girasol — a transparent blue white with a red play of color.

Harlequin — a variety in which the patches of color are small and angular but evenly distributed.

Hyalite — a transparent and colorless opal. It occurs in botryoidal masses.

Hydrophane — a variety that becomes transparent and sometimes shows a play of color when wet.

Lechosos — a milky opal with red and green flashes.

Lluviznandos — raindrop opal, the color flashes are very small. Sometimes also called rain fire opal.

Mexican Water Opal — colorless opal with a play of color. Similar to gyrosol.

Moss Opal — a massive variety of common opal containing dendritic inclusions of manganese or iron oxides.

Pintas — common opal.

Quincite — rose-colored common opal.

Zeasite — a variety of opal showing deep green flashes of color.

SAN JUAN DEL RIO

To reach the opal mines near San Juan del Rio, Querétaro, turn northeast on Morelos Oriente Avenue in San Juan del Rio. Travel 7.4 miles to a yellow bridge. Six-tenths of a mile farther, turn left on a dirt road toward the mountains. Three miles more and you will arrive in Trinidad. You can travel this far in a modern car. All opal mines in the vicinity of Trinidad are known as the Trinidad opal mines.

The most easily accessible mine near Trinidad is La Carbonera. It is owned by Sr. Eugenio Ontiveros of Querétaro. (See Querétaro for address.) Permission is usually granted to visit the mine and collect.

La Carbonera is about two miles west of Trinidad, on the top of the mountain. The dump and scar of the huge open pit can be seen from the highway between San Juan and Querétaro. The road to the mine is rough and steep. A four-wheel drive or pickup vehicle is recommended though the trip can be negotiated with a stock car that has plenty of power and road clearance. Do not try it with a low car. Porfirio Nieto of Trinidad is a good guide to the area.

La Carbonera mine is worked principally by hand. The miners are paid 80 cents per day. Opal mining is hard, backbreaking work in a tough matrix on a sun-baked hill.

Opal and chalcedony are found in seams, vugs, and small pockets in the rhyolite matrix. A sledge hammer and geologist's pick are indispensable for breaking the hard matrix.

Three different types of opal can be found in the Trinidad area: common opal, hyalite, and precious opal.

Saturday is the day the miners take their opal to Trinidad. Buyers usually arrive early in the morning. All sales are final and on the basis of bargaining. Women and children sell most of the low grade opal from either shallow dishes or jars. In bargaining for opal in jars, you usually save a peso if you have something into which you can transfer the opal. Jars are hard to come by in Trinidad. The men usually hold the better quality opal which is generally sold by the piece.

There are two other easily accessible opal mines in this area. See the map for their location.

If you wish to stay in the area, the best hotels are in Tequisquiapan about 17 miles from San Juan del Rio, a spa with mineral baths. The town is delightful. The Hotel Tequisquiapan is charming. Adolfo Frias, the patron, speaks English. The food is good, the rooms clean, and there is a fine swimming pool with mineral water.

See the map for a loop trip from San Juan del Rio, Tequisquiapan, or Querétaro. The two opal mines shown are near a good, paved road. The mine north of Tequisquiapan, La Guadalupana, is controlled by Francisco Cabrera, Iturbide No. 16, San Juan del Rio, Querétaro. Sr. Cabrera has a shop at this address and stocks opal both cut and rough, many other Mexican materials, he cuts stones, and makes jewelry, also has some mineral specimens though the stock is not large.

Columbus' Thumb, a landmark to the north that can be seen for many miles along Highway 45, is near Bernal (see the map). It is a high volcanic neck and worth the few miles' drive to Bernal for a close look.

Queretaro Area

Excellent chalcedony has been found on the Cerro de la Granadas, near Granadas, Querétaro. The area is reached by taking the road to Galindo and Amealco. The road goes south from Highway 45, about five miles west of San Juan del Rio. From Amealco, go west on the road to Coroneo. About five miles from Amealco, turn north to Granadas. This is rough country and not recommended for stock cars.

The green obsidian with crystobalite (?) inclusions seen in shops around Querétaro comes from a deposit near Apaseo, Guanajuato, on Highway 45. The deposit is north of Apaseo, near Octopan, Guanajuato.

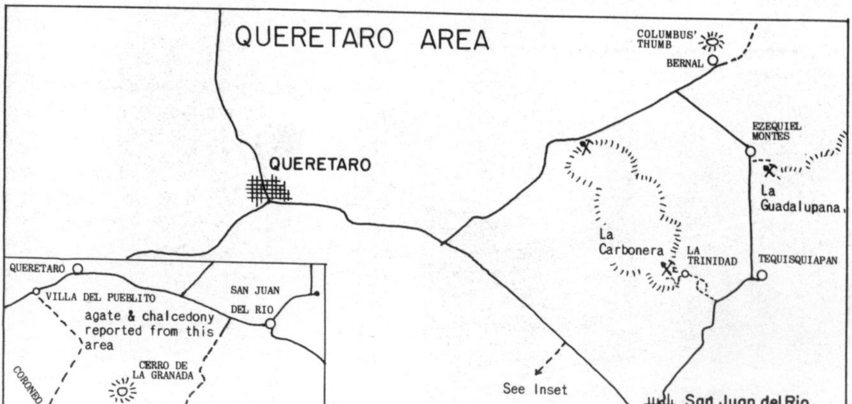

SECTION 3

Districts in this section are generally accessible from Highway 57, the eastern route from Piedras Negras to Mexico City via Monterrey and San Luis Potosí. Highway 57 joins Highway 45 at Querétaro.

AGUACHILE MOUNTAIN

This district is on the north rim of Aguachile Mountain, near Santo Domingo, Coahuila. It is reached by a road running northwest from Sabinas.

Fluorite was discovered here in 1955. It occurs as blue gray euhedral crystals. Beryllium was discovered in 1956. Bertrandite is the only beryllium mineral present. It occurs in the fluorite.

Minerals Reported

Aragonite	"Limonite"
Bertrandite	Muscovite, var. Sericite
Calcite	Orthoclase, var. Adularia
Fluorite	Quartz
Hematite	
Kaolinite	

SABINAS

Engineer José Delgado, Independencia No. 121, Sabinas, Coahuila, is in charge of the local office of the Comisión de Fomento Minero.

Sierra Mojada District

The Sierra Mojada District, a lead and silver producer, is in the extreme western part of Coahuila(s), a few miles from the Chihuahua(s) boundary. It is reached by road from Hermanas or Monclova, both on Highway 57.

The District was first discovered in 1878. Up to 1930 it had produced over five million tons of ore. The ore bodies are the manto type in limestone and dolomite. The ore is mainly oxidized lead, silver, and zinc minerals. There are also some copper silicates with high silver values.

Sierra Mojada District Mines

Aurora, La	Providencia
Buena Ventura	San Francisco
Encantada	San José
Esmeralda	San Salvador
Exploradora	Sultana, La
Fe, La	Tiro 10
Fortuna	Tiro 11
Galan Zona	Tiro Juarez
Jesús María	Veta Rica
Parena	Volcan Dolores

Minerals Reported

Anglesite	Erythrite
Antlerite	Feldspar group
Argentite	Galena
Azurite	Gypsum
Barite	Hematite
Calcite	"Limonite"
Celestite	Malachite
Cerargyrite	Pearceite
Cerussite	Proustite
Cerussite, Argentiferous	Psilomelane
Chalcocite	Pyrite
Chalcopyrite	Pyrolusite
Copper, Native	Quartz
Covellite	Silver, Native
Cuprite	Sphalerite
Descloizite	Strontianite
Dolomite	Sulfur, Native
Embolite	Wulfenite

Columbus' Thumb, a huge volcanic neck in the Queretaro opal country.

SALTILLO

The local office of the Comisión de Fomento Minero in Saltillo, Coahuila, is at Allende Sur No. 127. It is directed by Engineer Sebastian Romo.

Concepcion del Oro-Aranzazu District

Concepción del Oro is a silver-gold district. Aranzazú is a copper district. These districts are very close together in Zacatecas(s). They are reached by a road running southwest from Saltillo. The earliest mines were discovered about 1530.

The copper mineral klaprothite occurs here as irregularly rounded grains, long slender bladelike crystals, small euhedral crystals, and graphic intergrowths all of which are found within bornite. The klaprothite is less than 1/8 inch in size.

Concepcion del Oro-Aranzazu District Mines

Aranzazú	San Carlos
Cata Arroyo	

Minerals Reported

Andradite	Hornblende
Apatite	Idocrase (Vesuvianite)
Apatite, var. Collophane	Klaprothite
Azurite	Magnetite
Bornite	Malachite
Bournonite	Pyrite
Calcite	Pyroxene group
Chalcocite	Quartz
Chalcopyrite, Argentiferous	Scapolite
	Sphalerite
Chalcopyrite, Auriferous	Tenorite
	Tetrahedrite
Cuprite	Tetrahedrite, var. Freibergite
Diopside	
Epidote	Tremolite
Fluorapatite	Turquois
Galena	Wollastonite
Garnet group	Zoisite
Hematite	

MAZAPIL DISTRICT

The Mazapil District is in Zacatecas(s), a few miles west of Concepción del Oro.

Mazapil District Mines

Jesús María	Santa Isabel
Noche Buena	Todos Santos
San Carlos	

Minerals Reported

Arseniosiderite (pseudomorphous after Scorodite)	Chrysocolla
	Jamesonite
	Mazapilite
Azurite	Pharmacolite
Calcite	

MATEHUALA DISTRICT

The Matehuala District is located in the Sierra del Fraile, near Matehuala, San Luis Potosí. The Sierra del Fraile is composed of blue Mesozoic limestone overlain by shales. The largest mine in the District is the Dolores at the eastern base of the Sierra del Fraile.

Matehuala District Mines

Cobriza	Paza, La
Dolores	Santa María de la Paz
Esmeralda	

Minerals Reported

Actinolite	Hedenbergite
Andradite	Hornblende
Arsenopyrite, Auriferous	Idocrase (Vesuvianite)
Augite, var. Leucaugite (?)	Orthoclase
	Orthoclase, var. Adularia
Calcite	
Celestite	Prehnite
Chalcopyrite	Pyrite
Chlorite	Pyrite, Cupriferous
Epidote	Pyrrhotite
Fluorite	Quartz
Galena	Sphalerite
Galena, Argentiferous	Wollastonite
Grossularite	Zoisite

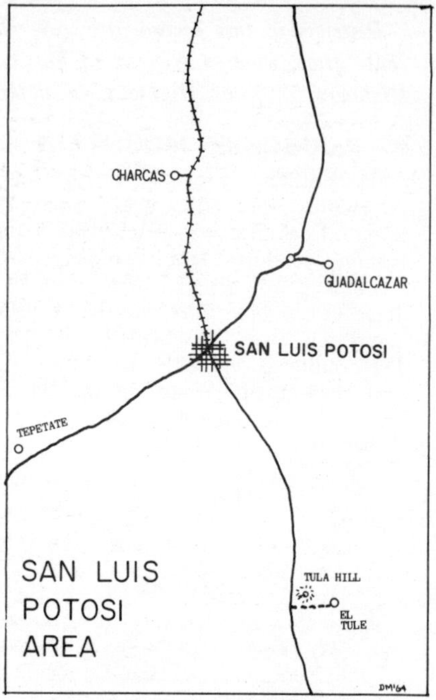

CHARCAS DISTRICT

The Charcas District in San Luis Potosí(s), was founded in 1574. Once worked for rich silver veins, it now produces lead, zinc, and copper. This District has long been famous for attractive white crystals of danburite. Colorful red cinnabar crystals up to one-quarter inch in length on white calcite are found here.

Tiro General is the principal mine in the District. La Bufa mine is also found here. The road to the District from Matehuala is very poor.

Minerals Reported

Calcite	Descloizite, var. Cuprodescloizite (ramirite or tritochorite)
Cinnabar	Pyrite
Danburite	Quartz, var. Carnelian

GUADALCAZAR DISTRICT

The Guadalcázar District is about 60 miles northeast of San Luis Potosí. It is a few miles east of Charco Blanco, the turnoff on Highway 57.

The dominant physiographic feature in the District is the Sierra de Guadalcázar. It trends westward and is flanked on the east by the Abrego basin, on the south by the Guadalcázar basin, and on the northwest by the Realejo basin.

This District was worked for mercury by Indians before its Spanish colonization between 1614 and 1620. Stone hammers used by the Indians have been found. Silver was found and worked as early as 1622. Later the District became the largest producer of mercury in Mexico. Gold has been washed intermittently from placer deposits for many years. Cassiterite (tin ore) was not recognized in the Guadalcázar District until 1937.

Geologically the District is composed of Cretaceous limestone which has been intruded by granite.

Vicente Mendez, both senior and junior, of Realejo, a village just north of Guadalcázar, are good guides to the local area.

Minerals Reported

Anatase
Andradite
Anglesite
Apatite
Argentite
Arsenopyrite
Axinite
Beryl
Beryl, var. Aquamarine
Bindheimite
Biotite
Brookite
Calomel
Cassiterite
Chalcopyrite
Chlorite
Cinnabar
Corundum
Danburite
Diopside
Dumortierite
Eglestonite (?)
Epidote
Fluorite
Galena
Goethite
Gold, Native
Grossularite
Gypsum
Hematite
Hematite, var. Specularite
Hornblende
Idocrase (Vesuvianite)
Ilmenite
Kermesite
"Limonite"
Livingstonite
Magnetite
Malachite
Manganese oxides
Matildite
Metacinnabar
Metacinnabar, var. Guadalcazarite
Metacinnabar, var. Onofrite (?)
Mimetite
Molybdenite
Monazite
Muscovite
Muscovite, var. Sericite
Oligoclase
Orthoclase
Polybasite
Powellite
Pyrite, Auriferous
Quartz
Quartz, Smoky
Rutile
Scheelite
Siderite
Sphalerite
Sphene (Titanite)
Spinel
Stephanite
Stibiconite
Stibnite
Svabite
Terlinguaite
Tetrahedrite
Topaz
Tourmaline, Velvet
Tremolite
Wolframite
Wollastonite
Xenotime
Zircon

SAN LUIS POTOSI

San Luis Potosí (sahn lu-IS po-to-SEE) is a major mining district of the interior plateau region with gold, silver, lead, copper, and fluorite mined in the local area. This refined, colonial city still has the charm of old Mexico.

The local Comisión de Fomento Minero office is at Independencia No. 80. Engineer Aurelio Martínez, is the director. The Comisión has a representative collection of minerals from the local area and San Luis Potosí(s). The collection includes azurite, barite, calcite, carnotite, cerussite, cinnabar, fluorite, galena, gypsum, molybdenite, opal var. hyalite, peridot, pyrolusite, pyrophyllite, quartz, and native sulfur.

Within a few hours' drive of San Luis Potosí it is possible to collect:

Calomel
Cinnabar
Enstatite, var. Bronzite
Fire Agate
Goslarite
Manganite
Mercury, Native
Opal
Psilomelane
Pyrolusite
Stibnite
Sulfur, Native
Topaz

The Pennsalt Chemical Company has a plant and office in San Luis Potosí. The company mines fluorite and owns the largest open pit fluorite mine in the world, the Consentida mine. In 1959 it produced 90,000 tons of fluorite ore. The ore is crushed at the mine and shipped by truck to San Luis Potosí. From there it is shipped by train to the United States for processing.

At the Consentida mine, the fluorite is massive and no crystals are found. The fluorite ore body is 1,800 feet long, 600 feet wide, and about 150 feet deep. The minerals found at the Consentida mine include fluorite, calcite, and pyrite. The fluorite is a replacement of limestone. On the west side of the Consentida mine the area is volcanic.

The mines are 33 miles from San Luis Potosí and consist of the open-pit Consentida mine and the Esperanza mine. The latter is a tunnel operation just being developed. You can collect as much fluorite as you want, and probably will be overloaded in a short time.

In the Esperanza mine the fluorite is also massive but occasionally a few small crystals are found. The fluorite in this mine originated from fault filling.

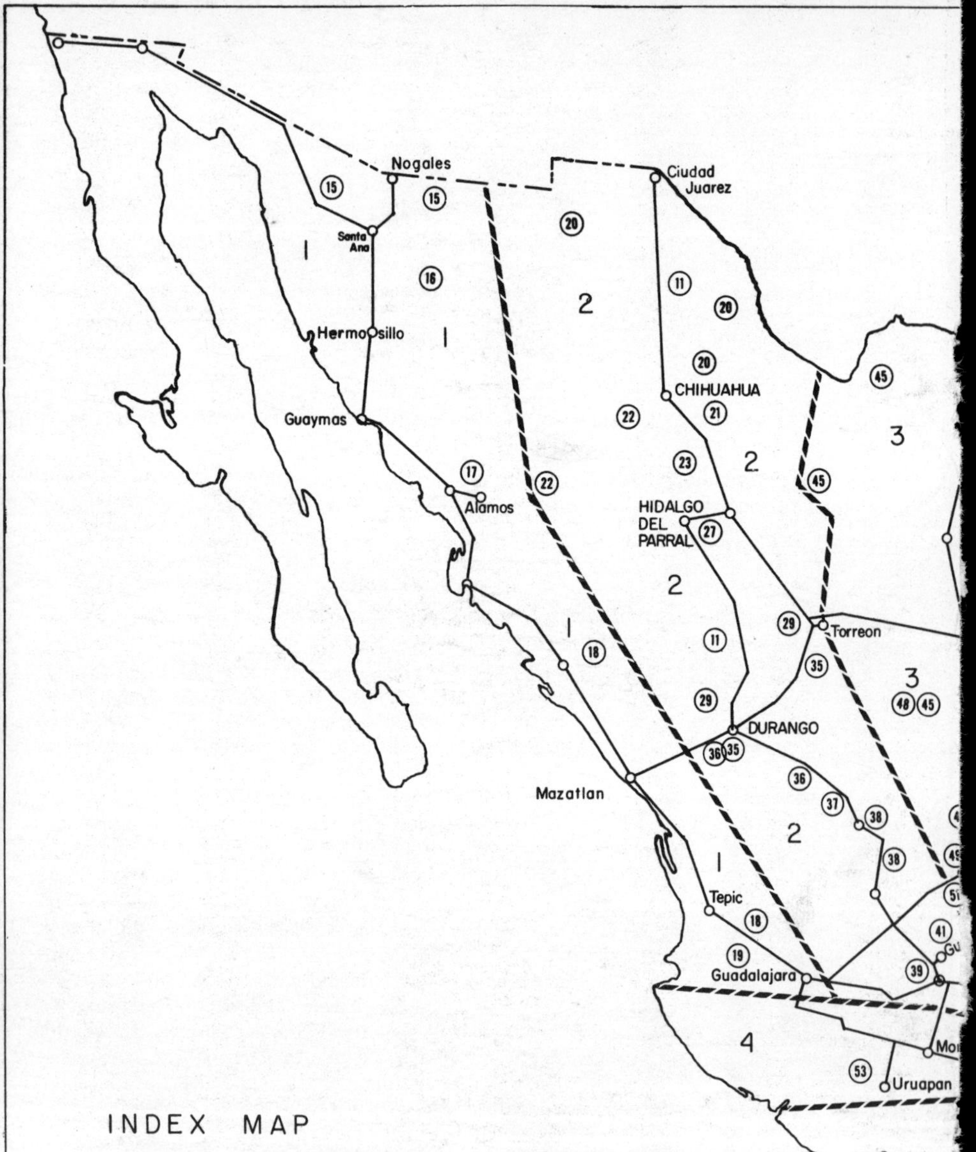

INDEX MAP to the GEMS and MINERALS of MEXICO

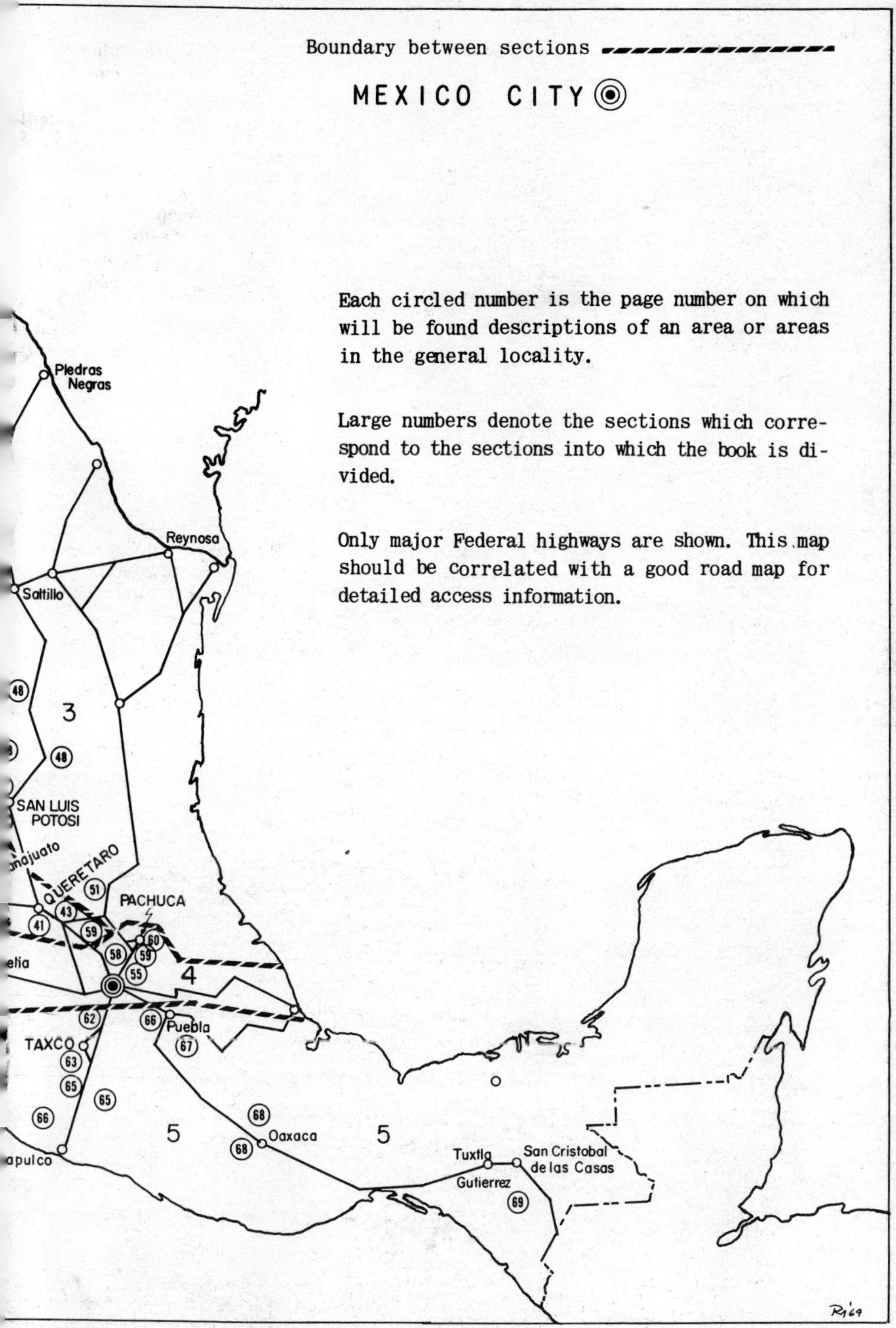

Each circled number is the page number on which will be found descriptions of an area or areas in the general locality.

Large numbers denote the sections which correspond to the sections into which the book is divided.

Only major Federal highways are shown. This map should be correlated with a good road map for detailed access information.

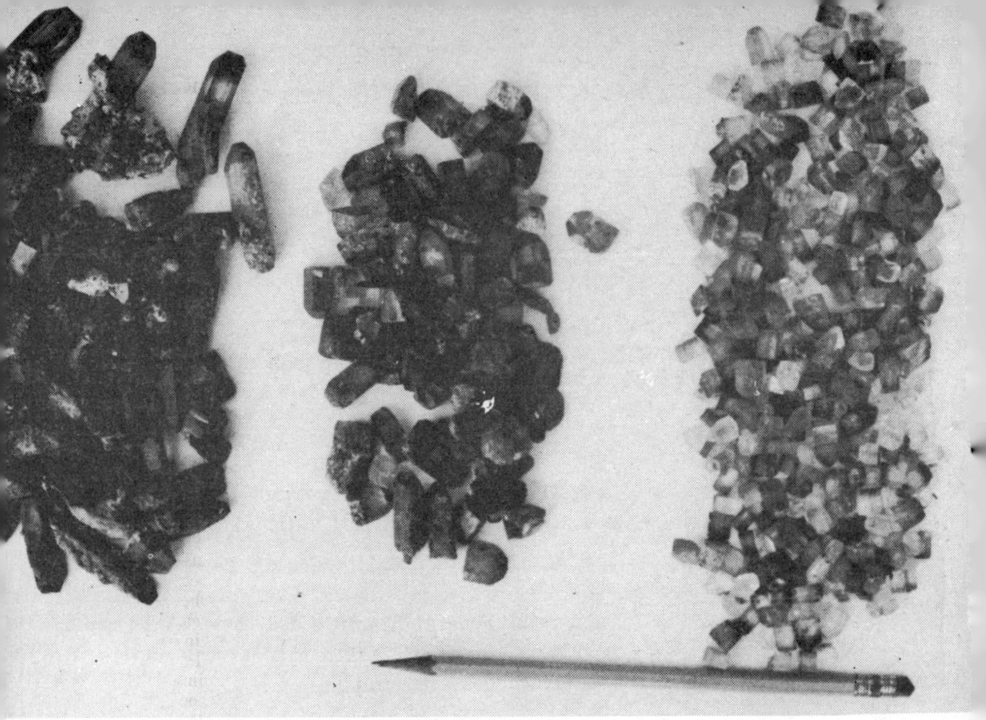

Sorted and graded topaz crystals from the mine near Tepetate, Mexico.

Golden topaz is found in the nearly vertical walls of the mine near Tepetate.

The City and State of San Luis Potosí, the Miners' Association, and civic bodies have recognized the economic value of visiting gem and mineral collectors. In the past they have cooperated by organizing field trips for collectors and offered generous cooperation. They have supplied guides, transportation, and information to both individuals and groups for a small fee. Contact Sr. Roberto Cuadra of the Miners' Association (Asociación de Mineros Potosino), Independencia No. 700, San Luis Potosí, about Operation Rockhound.

Lic. Felix Dauajara Torres, director of the local office of the Tourist Bureau, also may be helpful.

In the past, Operation Rockhound has arranged trips to the Tule Hill fire agate locality, Realejo (Guadalcazar) for minerals, Tepetate for topaz, Charcas (by train) for cinnabar, and several other localities.

San Luis Potosí is a delightful city and a good place to headquarter for trips into some of Mexico's finest collecting areas.

Tula Hill Fire Agate

In 1961 a fire agate locality was discovered about 45 miles south of San Luis Potosi at Tula Hill. The area is reached by turning off Highway 57 at Kilometer 359 onto the road to El Tule. The hill is just east of the main highway and north of the El Tule road.

Tepetate — Topaz

About 30 miles southwest of San Luis Potosi on Highway 80 is the village of Tepetate, near which are located the topaz mines. The mines are in rough back country and a guide is necessary unless you know specifically where you are going. Inocencio Ibarra Lopez of Tepetate may be available as a guide. Roads to the mines require truck type vehicles plus hiking.

Topaz is also found around Laurdes, southest of San Luis Potosi.

ZIMAPAN DISTRICT

This lead-silver district is near the town of Zimapan, Hidalgo. Mining began here in 1632. It has had a total silver production of about ten million dollars. The ore deposits are the replacement type in limestone. The ore is found in veins along faults in shaly limestone, fanglomerate, volcanics, and monzonite. Tin (cassiterite) veins have been discovered in rhyolite about 6.8 miles west of Zimapán.

Danburite from this District is occasionally found with a moving bubble inclusion.

Zimapan District Mines

Lomo del Toro	San Pascual
Los Balcones	

More minerals have been reported from this District than from any other in Mexico.

Minerals Reported

Actinolite	Melanterite
Adamite	Meneghinite
Alamosite	Mimetite
Albite	Minium
Analcime	Molybdenite
Andradite	Molybdite
Anglesite	Monticellite
Apatite	Muscovite
Argentite	Nepheline
Arsenopyrite	Olivenite
Augite	Olivine
Aurichalcite	Opal
Axinite	Orpiment
Azurite	Orthoclase
Barite	Orthoclase, var. Sanidine
Biotite	Plagioclase feldspars
Bornite	Plumbojarosite
Calcite	Pyrargyrite
Cassiterite	Pyrite
Cerussite	Pyromorphite
Chalcanthite	Pyrope
Chalcocite	Pyrrhotite
Chalcopyrite	Quartz
Chlorite	Quartz, var. Chalcedony
Chrysocolla	Realgar
Cinnabar	Rutile
Clinozoisite	Scapolite
Coccinite	Scorodite
Copper, Native	Serpentine
Danburite	Siderite
Diopside	Silver, Native
Dolomite	Smithsonite
Epidote	Sphalerite
Fluorite	Sphene
Galena	Spinel
Galena, Argentiferous	Spurrite
Gypsum	Stibnite
Hematite	Sulfur, Native
Hidalgoite	Tennantite
Hidalgoite, Ferrian	Tetrahedrite
Hornblende	Tourmaline
Hydromica	Tremolite
Hypersthene	Vanadinite
Iddingsite	Wollastonite
Idocrase (Vesuvianite)	Wulfenite
Jamesonite	Zeolite group
"Limonite"	Zinkenite
Loellingite	Zircon
Malachite	Zoisite
Manganese oxides	
Massicot	

Compiled by Simons and Mapes

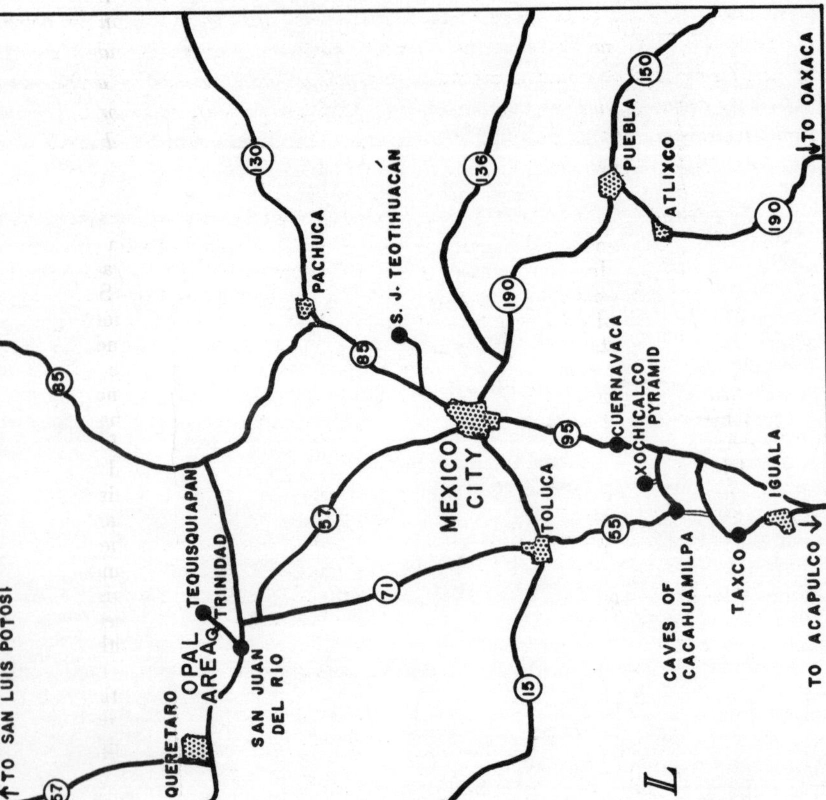
CENTRAL MEXICO
TO SAN LUIS POTOSI
TO GUADALAJARA
ZAMORA
CARAPAN
PARICUTIN
URUAPAN
PATZCUARO
MORELIA
LEON
SILAO
GUANAJUATO
IRAPUATO
SALAMANCA
CELAYA
QUERETARO
OPAL AREA
TEQUISQUIAPAN
TRINIDAD
SAN JUAN DEL RIO
PACHUCA
S. J. TEOTIHUACÁN
MEXICO CITY
TOLUCA
CUERNAVACA
XOCHICALCO PYRAMID
CAVES OF CACAHUAMILPA
TAXCO
IGUALA
TO ACAPULCO
PUEBLA
ATLIXCO
TO OAXACA
45
57
110
15
39
43
85
130
136
190
150
71
55
95

Section 4

Localities in this section are included in the broad band from Cape Corrientes on the west to Vera Cruz on the east. This unique geological region of volcanic rocks separates the northern plateau from the southern mountains and plains (see Physiography). It is not as productive mineralogically as the other areas, though exploration should yield much of interest to gem and mineral collectors. The main road from the west is Highway 15. To the east the region is roughly along Highway 136 from Mexico City to Vera Cruz.

PARICUTIN VOLCANO

Many old volcanoes have been observed re-erupting. In recorded history, however, only six volcanoes have been born and only one of these was actually observed from the start. It is therefore quite an interesting and unusual event to witness the creation of a new volcano.

On February 20, 1943, at 4:30 in the afternoon, near the village of Parícutin (pa-REE-koo-teen), in the State of Michoacan, Dionisio Pulido was tilling his cornfield. There had always been a hole on the ranch and Sr. Pulido and his family had used this area as a trash dump. Oddly enough, no matter how much they dumped into the hole it never seemed to fill up.

First a fissure passing through the hole opened and a small spiral of smoke issued forth followed by a fine, gray volcanic ash. It was this smoke that attracted the attention of Sr. Pulido and his neighbors. Tremors had been felt as early as February 5th, and subterranean noises continued to be heard until the smoke started to rise from the crack in the ground. Eyewitnesses said there was a smell of sulfur around the fissure. Sr. Pulido recalled later that he did not notice any heat in the ground around the crack.

The subterranean noise and smoke frightened the observers and, not realizing the importance of what they were witnessing, they fled to inform the local padre. The priest informed the Mexican government. Soon, scientists from all over the world were on hand to photograph and report a rare phenomenon — the actual birth of a volcano.

By 8 o'clock the following morning, the cone was about 30 feet high. The first lava started to flow within two days of the initial eruption, at a rate of about 15 feet per hour. Parícutin rose to a height of 500 feet in one week. In ten weeks it was 1,000 feet high. It grew to 1,797 feet above the surrounding terrain in eighteen months.

Parícutin eventually vomitted lava,

Paricutin, now dormant, looks down on the ruined church of San Juan which it destroyed and enveloped in lava. The size of the church and the flow are shown by the tiny figures in the foreground. One bell tower is all that remains standing.

The tremendous lava flow from Paricutin enveloped the entire village of San Juan, leaving only the church tower standing. Many other extinct volcanic cones can be seen in the background.

ash, cinders, bombs, and blocks bigger than railroad cars. Projectiles were hurled 4,120 feet into the air. The billowing ash plumes reached an altitude of 4 to 20 thousand feet. In the first seven months, the total material ejected was estimated to be two billion cubic yards. The temperature of the lava was 2,000 degrees Fahrenheit. The volcano covered thousands of acres with ash and changed the whole physical appearance of the region.

During the day the eruptions were set off in tones of gray. At night the fiery display continued with bursts of red, incandescent lava vivid against a black backdrop. The eruption continued without major interruption for nine years. Then it suddenly stopped on February 25, 1952.

According to scientists, it will not re-erupt and should, like so many of the

South of Guadalajara you drive into a country dotted with extinct volcanic cones like this one near Zamora.

other similar volcanoes of this area, remain extinct. During the first two and a half years of the eruption the lava was an olivine-bearing basaltic andesite, then a predominance of silica became apparent.

Two towns, Parícutin and San Juan Parangaricutiro (usually referred to as San Juan), were engulfed; more than 4,000 people had to flee their homes. The cone itself is three miles north of San Juan Parangaricutiro. The lava is very hard and sharp, and will soon reduce the best of shoes or boots to shreds. A single church steeple protrudes from the massive lava flow, to mark the village beneath.

To visit this famous volcano, turn off Highway 15 at Carapan (between Morelia and Zamora). Travel 23 miles south through Paracho (where guitars are handmade) and continue south for 12 miles, make a right turn just south of Capacuaro, and travel west. The small village of San Lorenzo, not shown on most maps, is two miles farther on.

In San Lorenzo a guide can be hired for the trip to the volcano. Do your bargaining for the guide service before hiring one. From San Lorenzo the dirt road extends 11 more miles. At the end of the road horses can be hired for about $2.00 per person. The ruins of San Juan Parangaricutiro are about two miles from the end of the road.

Benito Bernabe, a good guide, lives in San Lorenzo. The local inhabitants, mostly Tarascan Indians, speak little Spanish, their native tongue being an Indian dialect.

The tremendous amount of material ejected from Parícutin is plainly evident in the volcanic ash on the trail and in the depth of the beds exposed by water erosion, that can be seen from the road. A white spot near the cone is said by the guides still to be hot after 17 years!

The late Dr. William F. Foshag of the Smithsonian Institution wrote, "In my opinion, Parícutin is the greatest show on earth." He reported the natural pyrotechnics were so spectacular at one point that the spectators burst into applause!

Good accommodations are available in Uruapan.

MEXICO CITY

You probably have heard that the driving in Mexico City is hair-raising. Let me confirm the rumor. They don't play wrinkle-fender — they play for keeps! It is safer (?) and cheaper to take a taxi in town.

They tell the story in Mexico City about the motorist who knocked a man off a bicycle. As the man picked himself up from the street he shook his finger at the motorist and smiling, said, "*Cuidado!*" which means, "Be careful."

The Cabello Trailer Park, Hortensias No. 235, Mexico City, is very reasonable. It has a homelike atmosphere and is quite satisfactory.

Stopping at the embassy while visiting any foreign country is always a good rule. You may have your mail addressed to you in care of the embassy.

At the United States Embassy, Paseo de la Reforma and La Fragua, George M. Potter, a geological engineer, and Dr. Ralph L. Miller, a geologist with the U.S. Geological Survey, may be of considerable help in obtaining permission to visit different mining areas in Mexico.

The Compañía Minera de Peñoles, a Mexican mining company at 16th September No. 57, also may be of assistance in obtaining permission to visit its mines.

The American Smelting and Refining Company will probably give you a very cold reception in striking contrast to the warmth and courtesy displayed by others. Permission to visit its mines entails a lot of red tape and working through its New York City office.

A tremendous many-colored, natural, rough stone mosaic by Juan O'Gorman covers all four walls of the library of the University of Mexico.

National Pawn Shop

A visit to the National Pawn Shop, on the Zocalo across from the National Palace, should prove interesting. The pawn shop has quite a few emeralds mounted with diamonds. One four-carat emerald mounted with diamonds, was priced at $324.00. The commonest stone offered in the pawn shop is the alexandritelike synthetic corundum. These of course are sold as "genuine alexandrites." It is possible to find bargains here but most of the merchandise is actually new, much of it from stocks of bankrupt stores. A few good antiques are available but prices are relatively high. It is a good place to look but not to buy.

National University

The grounds of the National University of Mexico at University City (Ciudad Universitaria), a suburb of Mexico City, are expansive, well planned, and laid out to accommodate 60,000 students. The University, founded in 1551, is the oldest institution of higher learning in the western hemisphere. It is also the only one I know with a lava flow right on the campus. In fact it is built on a lava flow.

The National Geological Institute (Instituto de Geologia) is on the campus. Dr. Zoltan de Cserna, a geologist with the Institute, is thoroughly versed in Mexican

Slabbed travertine onyx covers this multi-storied wall on the Administration Building, University of Mexico.

geology and could prove helpful for suggestions. Dr. Eduardo Schmitter, mineralogist with the Institute, is in charge of its mineral collection. Professor Schmitter and his wife are most congenial and may be of assistance in the selection of localities to visit. Among other subjects, he is an authority on the mineralogy of the Oaxoca area.

Some idea of the large size of this malachite urn (with a malachite font and door in the background) is shown by the way it dwarfs the visitors to the Malachite Room in Chapultepec Castle.

Comisión de Fomento Minero

The Comisión de Fomento Minero is a Mexican government agency similar to the U.S. Bureau of Mines. It sometimes takes over mines that are bankrupt or running in the red. The Comisión has district offices in most of the major cities throughout Mexico. The engineers in charge of the various offices can help in supplying leads to collecting sites.

The general office in Mexico City is at Articulo 123 No. 37. Engineer Salvador Treviño is the general manager. Engineers Raymundo Llamas and Pascual Muñoz can be reached at the same address.

Chapultepec Park

In Chapultepec (cha-pool-tay-PAYK, which means grasshopper) Park, the grounds and gardens are arranged artistically. Many think they are comparable to the gardens of Versailles. Chapultepec Castle and Park are in the center of Mexico City. The castle houses the National Museum of History which is open daily from 9 a.m. to 5:30 p.m., and on Sunday from 10 a.m. to 2 p.m. Its collection of fine arts is supplemented by the rooms of Empress Carlotta and Emperor Maximilian, former President Diaz, and other historic figures.

One case in the Museum contains rare specimens of native silver crystals. Above all, find and photograph the rooms that house the superb, gigantic, Russian malachite bowls and the huge matched double doors of distinctive design in the same material. They are magnificent! A permit, usually 10 pesos, is required to photograph inside museums.

On the side of the hill as you leave the Museum, visit the caves or *grutas* (GROO-tahs) and be amused by the imitation stalactites and stalagmites made from plaster of Paris. It is well worth the four-cent admission.

Museum of Anthropology

The National Museum of Anthropology is just around the corner from the National Palace at Calle de la Moneda No. 13. This Museum contains one of the finest archaeological collections in the world. It is open every weekday from 10 a.m. to 8 p.m., and on Sunday from 10 a.m. to 2 p.m.

The Aztec Sacrifical Stone is just one of the huge stone carvings in the Museum of Anthropology, Mexico City.

A rough, natural stone mosaic, one of eight, decorating a fence of a private home on the Pedregal, Mexico City. This is the same technique used for the mosaic on the University library.

Here you may see the world-famous Aztec Calendar Stone and the Aztec Sacrificial Stone. The Calendar Stone weighs 40,000 to 60,000 pounds, depending upon which guidebook you read. (It doesn't really matter, because it probably will never be weighed.) It was carved from a single block of basalt three feet thick and twelve feet in diameter. The Aztecs had a very accurate calendric system employing a 365-day year based on astronomical observations.

The Sacrificial Stone has a small hole in the middle and a scupper to one side. The persons to be sacrificed were laid on the top of the stone and their hearts were cut out with a very sharp obsidian knife. Sacrificial knives also are on display.

There is a very extensive collection of jade artifacts. Not all of the material is genuine jade, as the ancients called any green stone "jade." There is also a magnificent display of gold work. Don't fail to see the obsidian artifacts, some of them magnificent examples of lapidary art. Then reflect for a moment that the immense stone carvings and delicate jade, obsidian, and gold were fashioned with primitive stone, wood, and bone tools. The pre-Columbian Mexican cultures did not have smelted metal for tools.

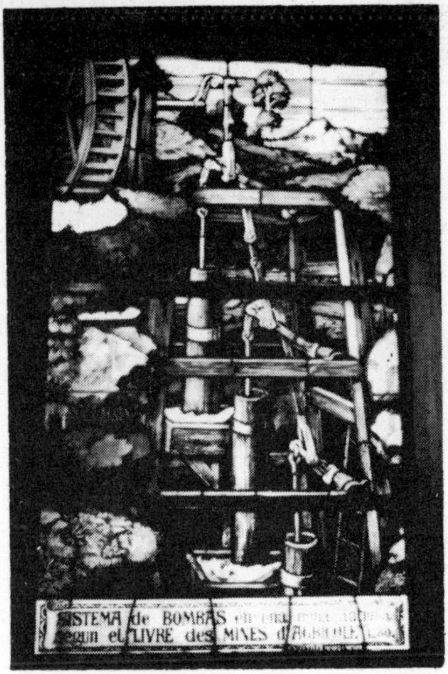

This stained glass window at the National Geological Museum ,Mexico City, shows mining in Germany centuries ago.

Geological Museum

A memorable time may be spent in the Geological Museum of the National Geological Institute. The Museum is located in the Geological Institution Building at Cipres No. 176, several blocks north of the railroad station. It is usually closed from 1 to 3 p.m. There you will find the finest collection of gems, minerals, fossils, and meteorites in Mexico. An added attraction is the exquisite stained glass windows brought from Germany. They depict mining operations in that country during the 1580's.

Ballet Folklórico de Mexico

Though it has nothing to do with gems or minerals, anyone visiting Mexico City should see the Ballet Folklórico de Mexico. It is not a ballet in the traditional sense, but a two-hour presentation of native dances from nearly every part of Mexico, presented in a colorful pageant at the Palacio de Bellas Artes. It is performed each Sunday morning, and tickets are available at hotels and travel agencies. It is well worth seeing.

You will also see the famous Tiffany glass curtain and an interior magnificently finished in travertine onyx from many parts of Mexico. There is also an exhibit of fine arts.

Tourist Bureau

The Mexican Tourist Bureau, Reforma 35, can be very helpful with general travel information and tourist attractions. It publishes maps and pamphlets on many subjects and areas in Mexico. The Bureau also maintains offices in most major cities in Mexico and in large centers in the United States.

The Pemex Travel Club, Juarez No. 97, has folders, maps, and travel information.

San Juan Teotihuacan

The Pyramids of Teotihuacán (teh-o-tee-wah-CAHN) and the Temple of Quetzalcoatl (ket-sahl-KOH-ahtl), the plumed serpent, are some of the major archaeological displays in Mexico. They are about 29 miles north of Mexico City. We found the area quite interesting and well worth the trip. The pyramids are believed to have been built by the Toltecs.

There is an interesting archaeological museum near the Pyramid of the Sun that contains a scale model of the vast, dead city. If you first study this model, it will help you to become oriented.

Around and on the pyramids, vendors sell carved objects of chatoyant (sheen) obsidian, both gold and silver. The gold

The Pyramid of the Sun is the focal point of the large archeological site just north of Mexico City. It is larger ,though not higher, than the famous pyramid in Egypt.

is said to represent the Pyramid of the Sun; the silver, the Pyramid of the Moon. Compare prices in the museum with those in the small market just outside, before buying. The museum prices are usually lower. The stand facing the walk in front of the museum usually has good values. The *patron* also has a lapidary workshop where obsidian is carved, in the village of San Juan Teotihuacan, which may be visited.

The pyramids have been excavated and restored. The base of the Pyramid of the Sun is larger than the largest pyramid in Egypt. A climb to the top is well worth the effort and affords a fine view of the surrounding excavations and the smaller Pyramid of the Moon.

Tula

Tula, the ancient capital of the Toltecs, is about 50 miles north of Mexico City. It is one of Mexico's major archeological sites. Since the completion of the new toll highway from Mexico City to San Juan del Rio, it is easily accessible over good roads. When traveling the highway, it is only a short detour and well worth the time and travel.

The area is one of sedimentary beds and material is mined for several cement plants.

The archeological site has a good museum. The pyramids and ball court are well preserved. It is here that you can see the *Atlantes,* huge carved basalt figures of warriors nearly 15 feet tall.

The big basalt figures at Tula.

PACHUCA-REAL DEL MONTE DISTRICT

Pachuca, about 80 miles north of Mexico City, is one of the most famous and oldest silver districts in the country. Silver was first discovered here in 1522 and subsequently was worked by Cortes' men. The mines come right down into the town. The most famous mine in the district is at Real del Monte about six miles east, over a good but winding road.

The Real del Monte mine was acquired by Romero de Terreros in 1739. By 1781 it had yielded 15 million pesos. In gratitude for loans and gifts to the King of Spain, Terreros was made Conde de Santa María Regla. Part of his fortune was also used to found *El Monte de Piedad,* (Mountain of Pity), now popularly known as the National Pawn Shop (see Mexico City).

Mines and dumps stud the mountains in the area. Many of them are easily accessible from the road. Though most of the material on the dumps is of little interest except to mineral collectors, good rhodonite of cutting quality has been reported from the mines at Real del Monte. Dogtooth amethyst crystals can be found on the dump of the Loretta mine and good micromounts on that of the Fresnillo mine. Both mines are at the first double-back curve on the road to Real de Monte, just at the edge of Pachuca.

George Zuniga, Gomes Farias 413, Pachuca, Hgo., is a good guide in the area if he is free. He teaches English and other subjects in the local school and collects personally. He is not a commercial guide or dealer.

Pachuca is not a tourist town and accommodations are limited. Hotel Noriega is good and clean though simple. The Club Español (Spanish Club) upstairs, about a half-block north of the hotel, serves reliable food.

Pachuca is on the northern tip of the central volcanic area and the geology is composed mainly of volcanic rocks. Pyrargyrite (ruby silver), common in most Mexican silver mines, has never been found here. The principal silver mineral is argentite.

Pachuca is reached from Mexico City by turning off Highway 85 at Colonia.

From the north, turn off Highway 85 about 12 miles south of Actopan. The turnoff is marked.

On Highway 85, between Actopan and the turnoff to Pachuca, to the east, are *Los Frailes* (The Friars), and on top of the mountain, *Los Organos* (The Organs). The former are unusual conical rocks that resemble marching friars. Los Organos are erosional features on the mountain top that resemble a pipe organ.

Pachuca-Real Del Monte District Mines

Acosta
Aranzazú
Arévalo
Buen Suceso
Camelia
Campechana
Ceciliana
Cruz, La
Dificultad
Dolores
Encino
Fraternidad, La
Jesús
Joya, La
Loreto
Morán
Navarro
Nuestra Señora de Guadalupe
Old Discovery
Palma, La
Parícutin
Peñol
Providencia
Quebradilla
Reunión, La
Rica, La
Rosario
Sabanilla
Sacramento
San Carlos
San Cayetano
San Felipe
San Francisco Javier
San Genaro
San Ignacio
San José
San Juan Pachuca
San Mateo
San Vicente
Santa Agueda
Santa Bárbara
Santa Cruz
Santa Teresa
Santo Domingo
Trinidad
Xacal
Zapatera, La

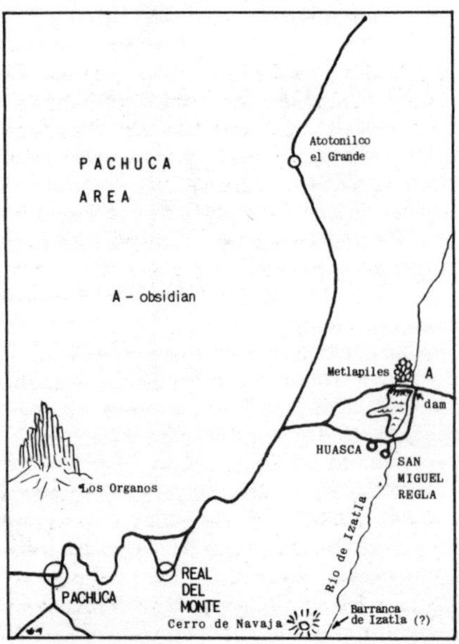

Minerals Reported

Acanthite
Albite
Allophane
Amesite
Anglesite
Anthophyllite
Antigorite
Aragonite
Argentite
Azurite
Barite
Biotite
Bornite
Braunite
Bustamite
Calcite
Cerargyrite
Cerussite
Chalcanthite
Chalcocite
Chalcopyrite
Chlorite
Chrysocolla
Clinochlore
Clinozoisite
Copper, Native
Coquimbite
Covellite
Cryptomelane
Dickite
Dolomite
Epidote
Fluorite
Fowlerite
Galena
Goethite
Gold, Native
Gypsum
Gypsum, var. selenite
Hematite
Hornblende
Jarosite
Johannsenite
Kaolinite
"Limonite"
Malachite
Manganite
Marcasite
Metavoltine
Miargyrite
Montmorillonite
Muscovite, var. Sericite
Neodigenite
Opal
Orthoclase, var. Adularia
Penninite
Plumbojarosite
Polybasite
Prehnite
Prochlorite
Proustite
Psilomelane
Pyrargyrite
Pyrite
Pyrolusite
Quartz
Quartz, var. Amethyst
Quartz, var. Chalcedony
Quartz, var. Jasper
Rhodochrosite
Rhodonite
Serpentine
Siderite
Silver, Native
Sphalerite
Stephanite
Sternbergite
Talc
Tetrahedrite
Tridymite
Wollastonite
Xonotlite (Xonaltite)

OBSIDIAN

Obsidian occurs in many places in Mexico. The fabled quarries of the Aztecs are in the mountains south of the village of Huasca, about 30 miles northeast of Pachuca. Huasca is not on most road maps but you can find San Miguel Regla on the AAA map. San Miguel Regla is an old hacienda that has been converted to an exclusive club for doctors and is about two miles east of Huasca. The road is rough but passable for stock cars.

Two old references to the obsidian mention the *Cerro de Navaja* (Hill of the Knife) and the *Barranca de Itzla* (Canyon of Glass). The Cerro de Navaja is about 10 or 15 miles due south of Huasca in the southwest corner of the

Municipio de Huasca. It is the highest peak on the horizon when looking south from Huasca.

The stream that runs through San Miguel Regla and through the columnar basalt locality (see below) is the *Rio de Itzla* (River of Glass). The Barranca de Itzla is presumably near the headwaters of this stream in the mountains to the south. The region is inaccessible to vehicles.

Obsidian, some of it gold sheen, can be collected in the Rio de Itzla and in the fields on each side of the stream.

Metlapiles

There is a massive columnar basalt formation called the *Metlapiles,* about three miles north of San Miguel Regla. It is difficult to see from the road because it is a deep canyon on the edge of a relatively flat plain.

Just north of San Miguel Regla is an artificial lake. A fair road runs north around the east shore of the lake. By following this road you eventually come to the dam that forms the lake. Turn left (west) on the road below the dam. At the far side of the dam is an open field with the Metlapiles on one side (west) and a cornfield on the other. You can drive right onto the field as it is uncultivated and hard, or park on the road. The top of the canyon is only a few steps away.

The spillway from the dam empties into the canyon on the west side. During the rainy season the water with its load of red dirt makes a colorful sight tumbling over the tops of the pentagonal and hexagonal basalt columns.

Collecting obsidian near the Metlapiles.

Packed columns of basalt in the Metlapiles.

Obsidian, about 20 per cent of it gold sheen, can be picked up in the field where you park. It is all in waterworn pebbles and boulders brought down by the stream in former years.

The bottom of the canyon can be reached by taking the road turning north on the west side of the canyon. The road drops down to the small village and hacienda at the mouth of the canyon. The floor of the canyon is part of the hacienda and permission must be obtained to enter.

This is a beautiful spot to visit, both for the obsidian and the unusual geological formation which rivals both Ireland's Giant's Causeway and California's Devils Postpile.

Though the Hacienda San Miguel Regla is a private club, it sometimes admits nonmembers for meals and lodging. Accommodations are luxurious, complete with swimming pool, golf course, and other recreational facilities.

This area is reached by taking the road from Pachuca to Real del Monte. Just past the summit of the grade, at a wye, take the left-hand road, bypassing Real del Monte. Continue down the canyon past old mine dumps and mill sites for about 10 or 12 miles to the turnoff marked "Huasca." From here on the road is strictly back-country road. It is passable but rocky and slow. If you turn left at the first major wye in this road, you will bypass Huasca and eventually arrive at the dam and Metlapiles. You can continue on around the east side of the lake to reach San Miguel Regla and Huasca.

SECTION 5

Most of the districts in this section are south of Mexico City and the broad volcanic belt (Section 4). A wide range of geology is encountered with a consequent wide variety of types of deposits. Little mineralogical information has been published on southern Mexico, but it is known that its potential is tremendous.

Cuernavaca

Cuernavaca, capital of Morelos(s), is the city of perpetual spring. Since the time of Cortes it has been the retreat of Mexico City's wealthy. On the east edge of the city, north of the railroad station, is the small Pyramid de Tempanzolco. This is a fine pyramid to climb for those who want to say they have climbed a pyramid in Mexico but are overawed by the huge structures at Teotihuacán.

The Cathedral, Borda Gardens, and Palace of Cortes are all worth a visit.

XOCHICALCO

Twenty-two miles southwest of Cuernavaca, off the road to the Cacahuamilpa Caves, are the ruins of Xochicalco (so-chee-CAL-co). At Xochicalco, Mexican archaeologists have excavated an unusually good ball court and pyramid.

The ball court has two stone rings, one on each side, through which a ball must be knocked. The ball game known as *tlachtli,* employed a hard rubber ball. The contestants had to play the game with their body, hips, and feet, and were not allowed the use of their hands. They played for keeps since it was customary to sacrifice the captain of the losing team. This pyramid is carved from basalt and is being restored. The ruins are believed to be of Toltec origin and about 1,000 years old.

The pyramid at Cuernavaca.

CAVES OF CACAHUAMILPA

Ninety-seven miles south of Mexico City, via Cuernavaca, are the Caves of Cacahuamilpa (cah-cah-wah-MEEL-pah). The caves were discovered in 1835. They were visited by Empress Carlotta, the wife of Maximilian, and many other famous figures of history.

Tours through the caves are conducted daily at regular intervals. For an extra fee a guide will take visitors in at any time. The grottoes are dimly lighted by electricity. Lectures are in Spanish. The caves are cool and the humidity is very high.

The domed ceilings rise to 150 feet or more. The effect is overwhelming! Watch your step — the floor and walks are very slippery, especially in the rainy season. Canes are sold outside the labyrinth and are well worth the few cents they cost, particularly if they keep you from slipping.

The caves are thought to extend for miles and have never been fully explored. Visitors are allowed to visit thirteen rooms which extend for about half a mile from the entrance.

Shale in the underlying strata is overlain by an impervious limestone cap. The

This pyramid at Xochicalco, near Cuernavaca, is being restored by archeologists.

caves are in the limestone with fantastically formed stalactites and stalagmites. The caves were formed by phreatic (ground water) circulation. The Cacahuamilpa Cave has blind solution cavities and solutional wall niches. Cacahuamilpa has a maximum width of 300 feet and maximum height of 180 feet.

The photographic opportunities in the cave are unlimited. We were permitted to wander at will underground without a guide. For more than four hours we took color photographs. We set our camera on a tripod and focused it by shining a flashlight on the subject. Setting the shutter on time, we opened it, and then walked around firing several electronic flashes. Then we closed the shutter. The pictures here show black and white reproductions of some of the color slides obtained.

There are five caves in this area, only three of which — Cacahuamilpa, Carlos Pecheco, and Cueva de la Estrella — may be entered. As the caves are a little off the main line of travel, the fame they merit still awaits them. In time, the world will discover them. Given proper illumination, they will rival the magnificence of Carlsbad Caverns.

TAXCO DISTRICT

Taxco (TAHS-co), Guerrero, located on a steep hillside, is a famous silver mining town. It is probably the oldest mining town in North America. Because it is picturesque, has good accommodations and weather, and is an easy few hours' drive from Mexico City, it is a popular tourist center.

One of the main attractions is the silver shops. There are over a hundred. Most of them cater strictly to tourists and the material sold is frequently shoddy and over-priced. The better shops, however, turn out exceptional work, and craftsmanship and design are both unusual and excellent.

Two shops worth a visit are Taller Borda on the Plaza next to the church, and Los Castillos about a block from the Plaza. The former specializes in silver and gold chess sets, among other things, and the work is exquisite. Tony Castillo's shop originated "married metals," a system of combining silver, brass, bronze, and other metals. These are worked into

The fantastic stalactites in the Caves of Cacahuamilpa would rival those of Carlsbad if properly illuminated.

A majestic stalagmite rises from the floor of the cave at Cacahumilpa. The domed ceiling is over 200 feet high.

This sterling silver knight from a chess set is an example of the superior silverwork that can be found in Taxco.

unusual flatware as well as jewelry. This shop also specializes in channel silver set with a variety of native stones.

The stones set in jewelry are generally rather poor. Most of them are used for ornamentation only, and a great deal of low grade malachite, fluorite, agate, jasper, some turquois, and similar materials is used. The value of good pieces

Brilliantly chatoyant blue tigereye and a small piece of nephrite jade (?), lower right, which the owner claimed were from the mountains near Acapulco.

is usually in the craftsmanship, not the stones.

Good stones can be purchased in Taxco, however. If you are looking for exceptional Mexican opal, contact the *patron* (owner) and make known your wants. He will usually take you into his office and bring out his valuable stones — opal, amethyst, jade (not native), Colombian emeralds, native and Brazilian aquamarine and topaz. The prices will not be low, but a little bargaining may help.

Not all the stores have their own workshops. When they do, they are usually very glad to let you visit and watch the workmen. It is a liberal education to see them turn out beautiful pieces with the simplest tools.

There are several shops in a group along the highway below the town. They all display the usual line of tourist material but all of them have good pieces if you are selective. Most of the faceted stones you will see are synthetic. For some reason, synthetic alexandritelike corundum seems to predominate, but you will find many other colors, too. Be very careful when buying faceted stones.

There are several excellent, though not low priced, hotels in Taxco.

The Taxco District was originally discovered about 1549. The first mining by Europeans in the western hemisphere was done here. Its total production has amounted to more than 8,820,000 pounds of silver.

The initial period of important mining in this District lasted from 1570-1585. The second major period of production was during the middle of the 17th century when the La Borda brothers were actively mining. An earthquake in 1802 destroyed the waterpower sources and nearly closed the mining operations. At the beginning of World War II, mining commenced again and production reached about 1,000 tons of ore per day. There are many small mines in the mountains around Taxco.

Early in the 16th century and again briefly in 1943 tin placers were worked in the Taxco District.

The main ore bodies, mostly veins, are found in limestone which overlies shale and schist. The veins dip steeply, strike roughly north and south, and usually outcrop on the surface. The workable veins vary in thickness from three feet to several yards and are mined by horizontal adits or tunnels. The ore veins are vertically zoned with quartz-calcite-silver veins above those of quartz-pyrite-sphalerite-galena. As is frequently true with most ore deposits, the richest ore in the Taxco District is found at the intersection of two veins.

The primary ore minerals are argentite, galena, and sphalerite, with smaller amounts of chalcopyrite. Pyrite, quartz, and calcite are the principal gangue minerals. The veins, besides being mined for silver, lead, and zinc, also contain a small amount of gold.

There is a fluorite mine about 10 miles north of Taxco, down in the canyon. It is mined primarily for fluorite as ore but some of the best purple material is cut into cabochons or used for inlays in Taxco.

Minerals Reported

Aikinite	Pyrargyrite
Argentite	Pyrite
Calcite	Quartz
Cassiterite	Quartz, var. Amethyst
Cerargyrite	Silver, Native
Chalcopyrite	Sphalerite
Fluorite	Stibiconite
Galena	

HUAHUAXTLA DISTRICT

This district is 17.7 miles south of Taxco, or 5.3 miles by road southwest of Puente Campuzano, Guerrero. This famous mercury district was discovered in 1923. Gallagher (1948) reports, "It is said that the early operators shoveled Montroydite crystals an inch long into their retorts by the sackfull." The rare minerals found in this district include calomel, eglestonite, montroydite, mosesite, and terlinguaite.

Minerals Reported

Calcite	Metacinnabar
Calomel	Montroydite
Cinnabar	Mosesite
Eglestonite	Pyrite
Gypsum	Stibnite
Marcasite	Terlinguaite
Mercury, Native	

HUITZUCO DISTRICT

The Huitzuco District in Guerrero, east of Iguala, was discovered in 1869. Mercury was first found there in 1873. Since then this District has had a total production of about 90,000 flasks. A flask holds 76 pounds of mercury.

After the initial period of mercury mineralization, heat and water formed

The "Normans and the English" in gleaming brass and silver lined up for battle on a wood and silver chessboard in Taxco.

mud geysers over a large area. Ores near the surface filled these extinct mud geysers with the secondary mercury minerals, stibiconite and cinnabar. These minerals can be found in the gravel and mud filling the vents of the geysers.

The deep deposits are composed of stockworks in brecciated blocks, veins, and replacements in limestone. The dominant rock in the District is limestone. As one goes more deeply into the workings, a greater amount of stibnite (antimony sulfide) is encountered. This probably indicates that the lower limit of the mercury mineralization is being approached. Stibnite, livingstonite, and sulfur occur in limestone, gypsum, and dolomite through a developed vertical range of 850 feet.

La Cruz is the principal mine in the Huitzuco District.

Minerals Reported

Aikinite	Metacinnabar, var. Guadalcazarite
Antimony oxides	Metacinnabar, var. Onofrite
Calcite	Opal
Cinnabar	Pyrite
Dolomite	Stibiconite
Fluorite	Stibnite
Graphite	Sulfur, Native
Gypsum	Terlinguaite
Gypsum, var Selenite	Tiemannite
Kermesite	Valentinite
Livingstonite	
Metacinnabar	

Rio Balsas Region

South of Iguala, Guerrero, the Rio Balsas and Rio Mexcala cut a huge canyon from east to west. The mountains between these two rivers and the Pacific Ocean embrace a vast region that is virtually unexplored. Gold has come from here since pre-Columbian times. Roads into the mountains are few and penetrate only a short distance; they are rugged and covered with growth. The only means of transportation is by horse or on foot.

Indians from the interior are reported to have brought out small quantities of jade (nephrite), emerald, amethyst, and other minerals. None of the reports can be confirmed, but there is little doubt that this great unexplored region is a potential treasure house. Someone with the time and knowledge will uncover its riches some day.

PUEBLA

Puebla, Puebla, about 90 miles east of Mexico City, is one of the country's most historic towns. It was here that the Mexicans defeated the French on May 5, 1862. This was the famous *Cinco de Mayo* celebrated as a national holiday and commemorated on street signs in nearly every city and village. The city's primary interest to the rock enthusiast is as the center of the onyx trade.

The onyx, seen all over Mexico, is actually travertine, not true onyx. Not all of it comes from around Puebla. The city is the heart of the carving trade. You will find carvings offered all over town. There are many *talleres* (shops). All of them are glad to admit visitors to watch the craftsmen. A good one to start with is Tesoros Artisticos de Onyx, 18 Poniente 103. They speak English.

Just up the street a few doors is a good tile factory you may visit, too. A few doors the other way, the famous Hidden Convent is worth a visit. Be sure to visit the Cathedral and see the

The onyx stores of Puebla carry a wide variety of carved, inlaid and fabricated artifacts as shown by George and Alice Sosa in their store. The checkerboard of inlaid onyx weighs about 75 pounds. The large Choc Mool (foreground)is a replica of an ancient piece, carved in onyx.

A shelf of onyx bookends.

onyx altar and fonts. The building was started in 1562 and completed in 1649.

Onyx is shipped into Puebla from all over Mexico. The main sources are the deposits near Tehuacán and Oaxaca and the local deposit near Tecali. So famous has the onyx from Tecali become that it is frequently called "tecali."

Tecali

Tecali is a small village southeast of Puebla. It can be reached in an hour's drive by turning off the main Puebla-Vera Cruz route, Highway 150, at Tepeaca. From Tepeaca to Tecali there is a good gravel road.

There are several small onyx works in Tecali. A good place to start is the Taller Ramon Tellez, 2 Sur No. 12. It is near the southwest corner of the Plaza. The shop has a good variety of excellent Mexican onyx carvings. The prices are considerably lower than Puebla's and much lower than Mexico City's. A little bargaining for large pieces helps.

Sr. Tellez' workshop is on the street on the opposite side of the Plaza, about two blocks south. You are welcome to watch the craftsmen work the onyx. Except for a small, gasoline driven diamond saw and one 100-grit wheel, the work is done entirely by hand.

The Tecali onyx quarries are in the mountains south of the town. It takes considerable persuasion but it is possible to obtain a guide to take you to the deposits. The roads are very poor and impassable in wet weather. Rough onyx can be purchased in the town in nearly any quantity you can carry. Sr. Tellez has a stock of onyx from the Tehuacan, Oaxaca, and Tecali quarries.

It is also worth a visit to the church

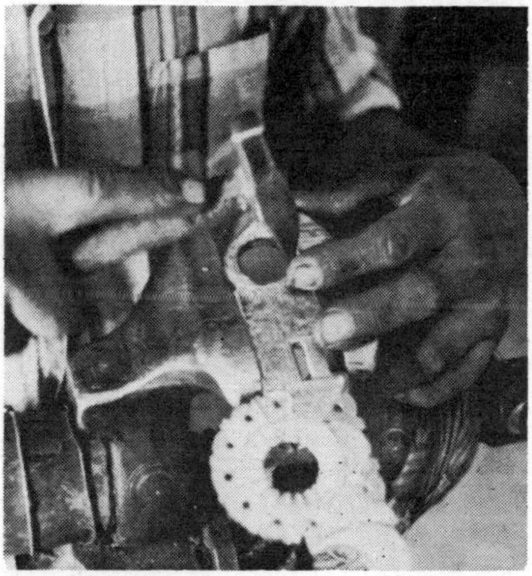

Hammering silver into a recess to decorate a bridle bit, one of the popular silver-steel objects made in the little town of Amozoc, near Puebla.

A selection of onyx necklaces made entirely by hand by the onyx carvers of Puebla.

to see the new altar at one of the stations of the cross. It is very modern and was made as a cooperative effort by the local craftsmen. Onyx from all over Mexico was used in its construction.

Amozoc

On the way to Tecali, stop off at the highway shops in Amozoc, a small village on Highway 150 about 10 miles from Puebla. This is the town where the famous steel-silver work is done. Most of the articles made are tack for horsemen — bits, stirrups, saddle decorations, buckles, etc. You can get other things, though, and it is worth the time and effort to see the workmen inlay the steel with good Mexican silver.

OAXACA

A vast, relatively unexplored region in southern Mexico awaits future prospecting by gem and mineral hobbyists. Little is known about the area mineralogically except isolated reports on small areas. Dr. Eduardo Schmitter, head of the Mineralogy Department, University of Mexico, has been working in the Oaxaca region. His enthusiasm for the area is intense. Many beautiful crystallized mineral specimens have been obtained from areas as yet completely inaccessible except by burro or horseback. He states that the region is full of unexplored pegmatites and other mineralizations of nearly every type.

Few collectors have visited southern Mexico except to view the unusual archeological sites of which there are many. The two most famous are Monte Albán and Mitla. Oaxaca is the center of widespread archeological activity by many Mexican and United States organizations.

We could obtain no specific information on collecting in this area but it is one that obviously will repay the ardent prospector who has time to spend. In any case, it is worth a visit for the archeological wonders alone.

A few references to various minerals in the Oaxaca region wil be found in the *List of Miscellaneous Minerals and Localities.*

TELIXTLAHUACA DISTRICT

The Telixtlahuaca District is in the Municipio of San Francisco Telixtlahuaca, Oaxaca. The District is just east of Highway 190, about 40 miles north of Oaxaca.

Minerals Reported

Albite	Muscovite
Albite, var. Cleavelandite	Muscovite, var. Sericite
Beryl	Orthoclase
Biotite	Quartz
Columbite	Spodumene
Fergusonite	Tourmaline
Ilmenite	Vermiculite
Kaolin group	

CHIAPAS

A vast and virtually unknown area, at least as far as gem and mineral potential is concerned, lies south of Oaxaca in Chiapas and along the border of Guatamala. This area has long been famous for it archeological wonders. The famous Mayan sites at Palenque and Bonampak which yeilded many stone artifacts, including jade pieces, are in this area. (Replicas of some of the artifacts and tombs can be seen at the Museum of Anthropology in Mexico City.) Both of these sites are in remote jungle areas and accessible only by special transportation.

Amber

The most famous of the gem materials reported from here is amber. Viking Orm, who has lived in San Cristobal de las Casas and explored the region extensively, stated in a letter, "It comes in chunks as big as your head and bigger. The deposit is on a steep fault scarp and is exposed for 1000 meters. The amber is found in veins or horizontal beds all along the face . . . and in the talus at the foot of the cliff . . . some of the amber is reddish, other clear, and some is yellow, all of it is transparent to translucent . . . the amber occurs in soft limestone and with it are very perfect fossils of fish, sea shells, mother-of-pearl, shark's teeth, etc."

Mr. Orm states the area is "full of rivers and the rivers are full of gemstones" such as agate and similar materials. Much of the area was placered for gold in the pre-Columbian era and Mr. Orm still works the placers successfully.

Tuxtla Gutierrez is the capital and largest city of Chiapas, but is located in a low tropical area. Headquarters for most prospectors is at San Cristobal de las Casas, an historic old city founded by followers of Cortes, which is at an elevation of about 7000 feet and quite comfortable in spite of the tropical location. This area is potentially one of the richest in Mexico from a collector's standpoint but is, as yet, almost totally unexplored.

MISCELLANEOUS MINERALS AND LOCATIONS

The following abbreviations are used in this section:

Am. Min. = *American Mineralogist.*

S.D.S.N.H. = data from locality labels in the collection of the San Diego Society of Natural History.

D.S. = Dana's *System of Mineralogy*, seventh edition.

D.T. = Dana's *Textbook of Mineralogy*, fourth edition.

Econ. Geol. = *Economic Geology.*

Gonzalez Reyna = *Memoria geologico-minera del Estado de Chihuahua,* (minerales metalicos), 280 p., XXth International Geological Congress, 1956, by Jenaro Gonzalez Reyna.

Nat. Geol. Mus. = data from locality labels in the collection in the National Geological Museum in Mexico City.

Rowland Collection = data from locality labels in the collection of Robert W. Rowland, Curator of Mineralogy, San Diego Museum of Natural History.

S.W.C. = Scott Williams Mineral Company, Inc., catalogue, copyrighted 1961.

Other references will be found in the Selected References section.

The numbers in parentheses after each reference below refer to the section of the Index Map in which it is located.

Alabandite

Preciosa mine, Cerro Tlachiaque, Puebla, Puebla. D.S., vol. 1, p. 208. (5)

Tetela de Ocampo, Puebla. Nat. Geol. Mus. (5)

Almandite

Rancho Corralitos, Municipio of Chihuahua, Chihuahua. Gonzalez Reyna, 1956, p. 278. (2)

Amber

Municipio de Ojo Caliente, Zacatecas. The owner is Gill Villalpando Ramos. Esquivel and Zamora, 1958, p. 89. (2)

From the alluvium near the Huitapan River, Simojovel District, Chiapas. This amber is red to yellow to black. Sinkankas, 1959, p. 602. (5)

From near San Cristóbal de Las Casas, Chiapas. Data supplied by Viking Orm. (5)

Andalusite, var. Chiastolite

Valle de Bravo District, Ixtapan del Oro, Mexico(s). Sinkankas, 1959, p. 478. (4)

Anglesite

Villa Ahumada, Chihuahua. Nat. Geol. Mus. (2)

Antimony, Native

Arechuybo, Chihuahua. Nat. Geol. Mus. (2)

Apatite

Mercedes mine, Sabinas Hidalgo, Nuevo Leon. Am. Min., vol. 37, p. 180. (3)

Apatite, var. Collophane

Astillero, near Mapimí, Durango. S.W.C., p. 6. (2)

Argentite

Taviche District, near Ocotlan, Oaxaca. (5)

Ario de Rosales District, 23 miles southeast of Ario de Rosales, Michoacan. (3)

Argentite, var. Jalpaite

Jalpa, Zacatecas. D.T., p. 418. (2)

Augite

La Panchita mine, Oaxaca, Oaxaca. Nat. Geol. Mus. (5)

Beryl

Sahuaripa, Sonora. Blue color. S.W.C., p. 10. (1)

Birnessite

From near the City of Zacatecas, Zacatecas. Am. Min., vol. 47, p. 790. (2)

Bismuthinite

La Gloria mine, Navajoa, Sonora. S.W.C., p. 11. (1)

Bismutite

Durango, Durango. D.S., vol. 1, p. 601. (2)

Santa Catarina mine, Sierra de Santa Rosa, near Guanajuato, Guanajuato. D.S., vol. 1, p. 278-279. (2)

Bromyrite

Carrizo mine, Municipio of Julimes, Chihuahua. Gonzalez Reyna, p. 273. (2)

Plateros District, Zacatecas. D.S., vol. 2, p. 14. (2)

Parrilla, Durango. D.S., vol. 2, p. 14. (2)

Brucite

Sombrerete, Zacatecas. (2)

Bustamite

Tetela de Xonotla, Puebla. D.T., p. 565. (5)

Calcite

Creel, Chihuahua. S.W.C., p. 14. (2)

Santo Domingo, Chihuahua. Nat. Geol. Mus. (2)

Carnotite

San Carlos, Chihuahua. Nininger, 1954, p. 32. (2)

Cassiterite

Coneto de Comonfort, Durango. Kunz, 1902, p. 58. (2)

Los Vallecitos mine, Teocaltiche, Jalisco. Foshag and Fries, 1942, p. 160. (1)

Cavillo tin district, Cavillo, Aguascalientes. (2)

Dolores Hidalgo tin district, 15 miles west of Dolores Hidalgo, on the road to Guanajuato, Guanajuato. (2)

Sain Alto District, 15 miles south of Sain Alto, Sierra de Chapultepec, Zacatecas. (2)

Celadonite

Cerro del Chanate, Municipio of Meoqui, Chihuahua. Gonzalez Reyna, p. 278. (2)

Cerussite

Aurora mine, Aldama, Chihuahua. Nat. Geol. Mus. (2)

Chalcanthite

Santo Nino mine, Municipio and Mral. of Batopilas, Chihuahua. Gonzalez Reyna, p. 280. (2)

Cobaltite

Cosala, Sinaloa. D.S., vol. 1, p. 446. (1)

Coccinite

Casas Viejas, near San José Iturbide, Guanajuato. D.S., vol. 2, p. 42. First found in Mexico. (2)

Cocinerite

From the 1100 foot level of the Cocinera mine, Ramos, San Luis Potosí. Doubtful species, needs confirmation. D.S., vol. 1, p. 173. (3)

Cosalite

Cosala, Sinaloa, D.S., vol. 1, p. 446. (1)

Cristobalite

Cerro San Cristóbal, near Pachuca, Hidalgo. D.T., p. 475. (3)

Cuprite

Cocinera mine, Ramos, San Luis Potosí. Am. Min., vol. 4, p. 146. (3)

Mapimí, Durango. Nat. Geol. Mus. (2)

Cuprotungstite

San Blas, Sinaloa. S.D.S.N.H. (1)

Daubreelite

Meteorite at Toluca, Mexico (s). D.S., vol. 1, p. 265. (4)

Iron meteorite, Bolson de Mapimí, Durango. D.S., vol. 1, p. 265. (2)

Descloizite

San Carlos or Benavides, Chihuahua. S.D.S.N.H. (2)

Diamond

No authenticated diamond locality has ever been discovered in the Republic of Mexico. Those that were reported were quartz, according to the description of the occurrence. Kunz, 1902, p. 92.

Durangite

Barranca tin mine, 18 miles northeast of Coneto de Comonfort and about 90 miles northeast of Durango, Durango. Found with cassiterite, hematite, and topaz. D.S., vol. 2, p. 830. (2)

Embolite

El Rosario mine, Municipio of Guadalupe y Calvo, Chihuahua. Gonzalez Reyna, p. 273. (2)

Fluorite

Monte Realejo, near Guadalcazar, San Luis Potosí. D.S., vol. 2, p. 34. (3)

Gold, Native

Altar District, Sierra del Bajio, Sonora. (1)

Bernal District, Bernal, Querétaro. (2)

El Oro, Mexico(s). (4)

Guadalupe y Calvo District, Guadalupe y Calvo, Chihuahua. (2)

Magdalena, Sonora. D.S., vol. 1, p. 93. (1)

Graphite

Santa María, Sonora. Upper Triassic coal beds 24 feet thick have been altered to graphite as the result of granitic dike metamorphism. D.S., vol. 1, p. 153. (1)

Campana Hill, near Hermosillo, Sonora. Aguilera, 1902, p. 498. (1)

Grossularite

From near Lake Jaco, Chihuahua. S.D.S.N.H. (2)

Rancho San Juan, near Xalostoc, Sierra Tlayacac, Morelos. In metamorphic limestone near its contact with igneous rocks. Associated with wollastonite and idocrase. Sinkankas, 1959, p. 292. (4)

Grossularite, var. Landerite

Also called rosolite and xalostocite. Rancho San Juan, near Xalostoc, Sierra Tlayacac, Morelos. In clear pink or rose-red dodecahedrons. Sinkankas, 1959, p. 292. (4)

Gypsum, var. Alabaster

Near Cruillas, Distrito Norte, Tamaulipas. Sinkankas, 1959, p. 566. (3)

Hematite

Cerro del Mercado, near Monclova, Coahuila. Econ. Geol., vol. 45, p. 120. (3)

Huebnerite

La Flaca mine, Rodeo, Durango. Nat. Geol. Mus. (2)

Idocrase (Vesuvianite)

Xalostoc, Morelos. Kunz, 1902, p. 58. (5)

From near Lake Jaco, Chihuahua. Rowland Collection. (2)

From the Sierra Tlayacac, Morelos. Am. Min., vol. 18, p. 25. (5)

Iodyrite

Municipio and Mral. of Cusihuiriachic, Chihuahua. Gonzalez Reyna, p. 273. (2)

Albarradon, near Mazapil, Zacatecas. First found in Mexico. D.S., vol. 2, p. 24. (2)

Jamesonite

La Gloria mine, Navajoa, Sonora. S.W.C., p. 35. (1)

Iguala, Guerrero. (5)

Johannsenite

Tetela de Ocampo, Puebla. Am. Min., vol. 23, p. 575. (5)

Kaolin

Santa María Coayuca, near Chignahuapan, Puebla. Aguilera, 1902, p. 503. (5)

Kaolinite

Barranca de Cobre (Copper Canyon), Municipio of Urique, Chihuahua. Gonzalez Reyna, p. 278. (2)

Kermesite

Arechuybo, Chihuahua. S.W.C., p. 36. (2)

Lead, Native

Jalapa, Veracruz. D.S., vol. 1, p. 103. (4)

Legrandite

Flor de Peña mine, Lampazos, Nuevo Leon. First found in Mexico. D.S., vol. 2, p. 959. (3)

Malachite

Cocinera mine, Ramos, San Luis Potosí. Am. Min., vol. 4, p. 146. (3)

Manganosiderite

Angangueo, Michoacan. Econ. Geol., vol. 20, p. 56. (4)

Marcasite

Chihuahua, Chihuahua. S.W.C., p. 40. (2)

Angangueo, Michoacan. Econ. Geol., vol. 20, p. 56. (4)

Massicot

From the volcano Popocatepetl, Mexico(s). D.S., vol. 1, p. 516. (4)

Perote, Zomelahuacan District, Veracruz. D.S., vol. 1, p. 517. (4)

Melaconite

Cocinera mine, Ramos, San Luis Potosí. Am. Min., vol. 4, p. 146. (3)

Metacinnabar

San Onofre, near Plateros, northwest of Zacatecas, Zacatecas. D.S., vol. 1, p. 216. (2)

Metacinnabar, var. Onofrite

San Onofre, near Plateros, northwest of Zacatecas, Zacatecas. D.S., vol. 1, p. 216. (2)

Minium

Santa Fe Mine, Bolanos, Jalisco. D.S., vol. 1, p. 518. (1)

Molybdenite

From near La Trinidad, Sahuaripa District, Sonora. D.S., vol. 1, p. 330. (1)

Nickel, Native

Toliman District, Querétaro. Aguilera, 1902, p. 505. (2)

Niter

Municipio of Ojinaga, Chihuahua. Gonzalez Reyna, p. 279. (2)

Obsidian

Pica de Orizaba, (Mt. Orizaba), Veracruz(s). Sinkankas, 1959, p. 509. (4)

Zacaultipan, 15 miles south of Tulancingo, Hidalgo. Sinkankas, 1959, p. 509. (4)

Obsidian, Chatoyant

Magdalena, Jalisco. Kunz, 1902, p. 84. (1)

Cerro de Navajas (Hill of Knives) on the Guajulate Estate, near Pachuca, Hidalgo. Kunz, 1902, p. 84. [See Pachuca.] (4)

Opal, Precious

Simpática mine, Hill of Ceja de Leon, Esperanza, ten leagues (30 miles) northwest of San Juan del Río, Querétaro. Kunz, 1902, p. 63. (2)

Rosario mine, Hill of Jurado, Esperanza, ten leagues (30 miles) northwest of San Juan del Río, Querétaro. Kunz, 1902, p. 63. (2)

Sisoguichic, Sierra de Tarahumara, Chihuahua. Foshag, 1953, p. 281. (2)

Tlaxiaco, Oaxaca. Foshag, 1953, p. 281. (5)

Barranca Agua Dulce, Tulancingo District, Hidalgo. Sinkankas, 1959, p. 128. (4)

Hacienda La Llave, San Juan del Río District, Querétaro. Sinkankas, 1959, p. 128. (2)

Hacienda Tesquisquiapán, San Juan del Río District, Querétaro. Sinkankas, 1959, p. 128. (2)

Purisima mine, Amelaco District, Querétaro. Sinkankas, 1959, p. 128. (2)

Hacienda La Jurada, Toliman District, Querétaro. Sinkankas, 1959, p. 128. (2)

Hacienda La Esperanza, Toliman District, Querétaro. Sinkankas, 1959, p. 128. (2)

Cerro Agustino, Maravatio District, Municipality of Maravatio, Michoacan. Sinkankas, 1959, p. 129. (4)

Opalized Wood

Fortuna mine, Monclova District, Sierra Mojada, Coahuila. Sinkankas, 1959, p. 125. (3)

Puebla de San Lucas, Texcoco District, Municipio de Chiautla, Mexico(s). Sinkankas, 1959, p. 129. (4)

Percylite

State of Sonora. Associated with gold and "limonite." First found in Mexico. D.S., vol. 2, p. 82. (2)

Petrified Wood

From near Mesquital, Durango. Sinkankas, 1959, p. 402. (2)

Salitera mine, Huichapan District, Hidalgo. Sinkankas, 1959, p. 402. (4)

Hacienda de Yextho, Huichapan District, Hidalgo. Sinkankas, 1959, p. 402. (4)

Sierra de las Silicates, near Tlaxcala, Tlaxcala. Sinkankas, 1959, p. 402. (4)

Polybasite

Taviche District, near Ocotlan, Oaxaca. (5)

Campania mine, Sultepec, Mexico(s). D.S., vol. 1, p. 360. (4)

Portlandite

Cerro de La Corona, Morelos, Coahuila. Am. Min., vol. 35, p. 614. (3)

Proustite

Taviche District, near Ocotlan, Oaxaca. (5)

"Psilomelane"

Numerous gem and mineral dealers are selling a black, submetallic material from Mexico under the name of "psilomelane." This attractive stone, usually tum-

bled, is also known as *Corona de la Plata* (Crown of Silver) and "black malachite." The latter names should not be used. Tests have shown that the material is a banded chalcedony with the black due to manganese oxides. It is reportedly found near Navajoa, Sonora. It has a hardness greater than six. A sodium carbonate bead turns blue green indicating the presence of manganese. (1)

Pyrargyrite

Ario de Rosales District, 23 miles southeast of Ario de Rosales, Michoacan. (4)

Taviche District, near Ocotlan, Oaxaca. (5)

Pyrite

Flor de Peña mine, Lampazos, Nuevo Leon. D.S., vol. 2, p. 959. (3)

Pyrope

Lake Jaco, Chihuahua. Kunz, 1902, p. 57. (2)

Bolson de Mapimí, Durango. Sinkankas, 1959, p. 292. (2)

Pyrrhotite, var. Troilite

Rancho Chupadero, Municipio of Hidalgo del Parral, Chihuahua. Gonzalez Reyna, p. 270. (2)

Quartz, Red

This red quartz, often referred to as "strawberry" quartz, was thought to have inclusions of rutile. Recent investigations (Sinkankas, 1962, p. 677) indicate the inclusions are fine needles of hematite. It reportedly is found near Uruachic, Chihuahua. (2)

Quartz, var. Agate

Teocaltiche District, Jalisco. Sinkankas, 1959, p. 402. (1)

Nazas River, near Rodeo, Durango. Sinkankas, 1959, p. 401. (2)

From west of Ciudad Camargo, Chihuahua. Sinkankas, 1959, p. 401. (2)

Tlalpujahua District, Tlalpujahua, Michoacan. Sinkankas, 1959, p. 402. (4)

Quartz, var. Amethyst

From near Culiacan, Sinaloa. Sinkankas, 1959, p. 401. (1)

Quartz, var. Chalcedony Geodes

From near Oputo, a small town on the De Bavispe River, about 25 miles southeast of Nacozari, Sonora. Sinkankas, 1959, p. 398. (1)

Quartz, var. Jasper

Tlalpujahua, Maravatío District, Michoacan. Sinkankas, 1959, p. 402. (4)

Espiritu Santo mine, Maravatío District, Michoacan. Sinkankas, 1959, p. 402.

Rhodochrosite

From near Magdalena, Sonora. S.W.C., p. 53. (2)

Espejeras, Tetela de Ocampo, Puebla. Nat. Geol. Mus. (5)

Imuris, Sonora. Nat. Geol. Mus. (5)

Sartorite

Mral. de Corralitos, Municipio of Casas Grandes, Chihuahua. Gonzalez Reyna, p. 271. (2)

Scheelite

Tecoripa District, 9.3 miles northwest of Tecoripa, Sonora. (1)

Siderite

Flor de Peña mine, Lampazos, Nuevo Leon. D.S., vol. 2, p. 959. (3)

Sillenite

Durango, Durango. Am. Min., vol. 28, p. 525. (2)

Silver, Native

Bernal District, Bernal, Querètaro. (2)

Mexico mine, El Oro, Coahuila. Am. Min., vol. 18, p. 297. (3)

Stephanite

Campania mine, Sultepec, Mexico(s). D.S., vol. 1, p. 360. (4)

Stromeyerite

Yoquivo mine, Yoquivo, Municipio of Ocampo Chihuahua. Gonzalez Reyna, p. 269. (2)

San Francisquito mine, Municipio of Chihuahua, Chihuahua. Gonzalez Reyna, p. 269. (2)

Rio Tinto Mexicano mine, Municipio of Chihuahua, Chihuahua. Gonzalez Reyna, p. 269. (2)

Sulfur, Native

Jaltipan, Veracruz. Sheehan, 1960, p. 101. (4)

Talc

San Gabriel mine, Municipio of Batopilas, Chihuahua. Gonzalez Reyna, p. 278. (2)

Tennantite

Teziutlan copper-zinc district, Teziutlan, Puebla. (5)

Thomsonite

Las Animas, Municipio of Aquiles Serdan, Chihuahua. Gonzalez Reyna, p. 278. (2)

Topaz

Hacienda Tepezate, Pinos, Zacatecas. Kunz, 1902, p. 58. (2)

Coneto de Comonfort, Durango. Kunz, 1902, p. 58. (2)

Mesa de Santa Cruz, near the city of San Luis Potosí, San Luis Potosí. Kunz, 1902, p. 58. (3)

Cerro del Nado, Tulancingo District, south of Apulco, Hidalgo. Sinkankas, 1959, p. 105. (4)

Sain Alto District, 15 miles south of Sain Alto, Sierra de Chapultepec, Zacatecas. (2)

From near Tepozan, Cerritos, San Luis Potosí. Sinkankas. 1959, p. 105. (3)

Cerro de la Cruz, Piños area, Zacatecas. Sinkankas, 1959, p. 105. (2)

Topaz deposits extend from the States of Zacatecas to San Luis Potosí. It is colorless to rich wine-yellow. Crystals from one to three inches in length are found and are sometimes doubly terminated. Kunz, 1902, p. 58.

Tourmaline, var. Schorl

Black "velvet" tourmaline from Santa Cruz, Sonora. S.W.C., p. 62. (1)

Travertine (Onyx)

Antigua Salines, Tehuacan District, Puebla. Sinkankas, 1959, p. 561. (5)

Tridymite

Cerro San Cristóbal, near Pachuca, Hidalgo. D.T., p. 475. (4)

Trona

Municipio of Julimes, Chihuahua. Gonzalez Reyna, p. 278. (2)

Ojo Caliente, Municipio of Casas Grandes, Chihuahua. Gonzalez Reyna, p. 278. (2)

Valentinite

Arechuybo, Chihuahua. S.W.C., p. 64. (2)

Vanadinite

San Carlos or Benavides, Chihuahua. Nat. Geol. Mus. (2)

Whitlockite

Mercedes mine, Sabinas Hidalgo, Nuevo Leon. Am. Min., vol. 37, p. 180. (3)

Wulfenite

Magdalena, Sonora. D.S., vol. 2, p. 1085. (1)

Aurora mine, Aldama, Chihuahua. Nat. Geol. Mus., Mexico City. (2)

Zircon

La Panchita mine, Oaxaca, Oaxaca. Nat. Geol. Mus. (5)

Sain Alto District, 15 miles south of Sain Alto, Sierra de Chapultepec, Zacatecas. (2)

SELECTED REFERENCES

Anonymous, 1960, **Caminos de Mexico,** 303 p., 5th edition, Goodrich Tire Company, 20 pesos.

Anonymous (issued annually) **Mexico and Central America,** American Automobile Association, Washintgon, D.C.

Barbour, Joyce, 1964, Mexican topaz and where it comes from, p. 22, 23, 61-63, January, **Gems & Minerals.**

Barbour, Joyce, 1964, How the Mexicans work calcite onyx, p. 14-18. June, **Gems & Minerals.**

Barbour, Joyce, 1964, Black opal at San Juan del Rio, p. 26-27, October, **Gems & Minerals.**

Barron, E. M. 1958, The Gem Minerals of Mexico, p. 4-16, April, **Lapidary Journal.**

Bassett, Allen M., 1956, Genesis and paragenesis of silicates and sulfides, Naica mine, Chihuahua, Mexico, XXth International Geological Congress, Mexico, D.F., in press.

Blake, William P., 1902, Notes on the mines and minerals of Guanajuato, Mexico, p. 216-223, vol. 32, **Transactions of the American Institute of Mining Engineers.**

Bonillas, Y. S., 1937, Geology of the Taxco mining district, Guerrero, Mexico (abstract), p. 200, vol. 32, **Economic Geology.**

Bretz, J. Harlen, 1955, Cavern-making in a part of the Mexican plateau, p. 364-375, vol. 63, **Journal of Geology.**

Cornwell, Mary, 1962, Mineral collecting assumes new dimension, mission to Mexico accomplished, (Operation Rockhound), p. 109-116, vol. 15, No. 3, June, **Earth Science.**

Cousen, Thomas W., 1956, Mexican fire opal, p. 16-17, 68-71, March, **Gems & Minerals.**

Dana, Edward Salisbury, 1932, **A textbook of mineralogy,** revised by William E. Ford, fourth edition, 851 p., John Wiley & Sons, Inc.

Dana, James D. and Dana, Edward S., vol. 1 (1944), vol. 2 (1951), vol. 3 (1962), **System of mineralogy,** seventh edition, rewritten by Charles Palache, Harry Berman and Clifford Frondel; John Wiley & Sons, Inc.

Edelen, A. W. and Lee, Herbert V., 1941, The Teziutlan copper-zinc deposit, Teziutlan, Puebla, Mexico, p. 314-323, vol. 144, **Transactions of the American Institute of Mining and Metallurgical Engineers.**

Esquivel M., J. and Zamora, S., 1958, **Informe sobre minerales no metalicos,** 226 p., Boletin 44, Consejo de Recursos No Renovables.

Foshag, William F., 1927, The selenite caves of Naica, Mexico, p. 252-256, vol. 12, No. 6, **American Mineralogist.**

Foshag, William F., 1928, Mineralogy and geology of Cerro Mercado, Durango, Mexico, 27 p., article 23, vol. 74, plates 1-4, No. 2768, **Proceedings of the United States National Museum.**

Foshag, William F., 1928, The minerals of Mexico, p. 21-26, **Explorations and field-work of the Smithsonian Institution in 1927,** publication number 2957.

Foshag, W. F., Further mineral collecting in Mexico, p. 33-40, **Explorations and field-work of the Smithsonian Institution in 1931,** publication number 3134.

Foshag, William F., 1934, The ore deposits of Los Lamentos, Chihuahua, Mexico, p. 330-345, vol. 29, No. 4, **Economic Geology.**

Foshag, W. F., 1935, Mineralogical investigations in Mexico, p 5-8, **Explorations and field-work of the Smithsonian Institution in 1934,** publication number 3300.

Foshag, William F., 1937, Carminite and associated minerals from Mapimi, Mexico, p. 479-484, vol. 22, **American Mineralogist.**

Foshag, W. F., 1940, Quest for gems and minerals in Mexico, p. 1-4, **Explorations and field-work of the Smithsonian Institution in 1939,** publication number 3586.

Foshag, William F., 1953, Mexican opal, p. 278-282, vol. 7, No. 9, Spring, **Gems and Gemology.**

Foshag, William F. and Fries, Carl, Jr., 1942, Tin deposits of the Republic of Mexico, p. 98-176, **U.S. Geological Survey Bulletin 935-C,** U.S. Government Printing Office, Washington, D.C.

Foshag, William F., and Gonzalez, Jenaro, 1956, Birth and development of Paricutin volcano, Mexico, p. 354-489, **United States Geological Survey Bulletin 965.**

Fowler, George M.; Hernon, Robert M., and Stone, Edwin A., 1944, The Taxco mining district, Guerrero, Mexico, (abstract), p. 93-94, vol. 39, **Economic Geology.**

Fries, Carl Jr. and Schmitter, Eduardo, 1948, Tin-bearing placers near Guadalcazar, State of San Luis Potosi, Mexico, p. 109-150, **U.S. Geological Survey Bullet 960-D.**

Gallagher, David, 1952, Geology of the quicksilver deposits of Canoas, Zacatecas, Mexico, p. 47-85, **U.S. Geological Survey Bulletin 975-B.**

Gallagher, David and Perez, Rafael, 1946, Geology of the Cuarenta mercury district, State of Durango, Mexico, p. 155-168, **U.S. Geological Survey Bulletin 946-F.**

Gallagher, David, and Perez, Rafael, 1948, Geology of the Huahuaxtla mercury district, State of Guerrero, Mexico, p. 149-173, **U.S. Geological Survey Bulletin 960-E.**

Gonzalez Reyna, Jenaro, 1956, **Riqueza minera y yacimientos minerales de Mexico,** third edition, 497 p., XXth International Geological Congress, Banco de Mexico, S.A. (In Spanish.)

Green, James A., 1944, Paricutin, the cornfield that grew a volcano, p. 129-156, vol. 85, No. 2, February, **National Geographic Magazine.**

Guiza, Reynaldo, 1956, The Guanajuato mining district, p. 123-128, XXth International Geological Congress guidebook for excursion A-2.

Hewitt, William Paxton, 1943, Geology and mineralization of the San Antonio mine, Santa Eulalia district, Chihuahua, Mexico, p. 173-204, vol. 54, part 1, **Bulletin of the Geological Society of America.**

Holmes, W. H., 1900, The obsidian mines of Hidalgo, Mexico, p. 405-416, vol. 2, **American Anthropologist.**

Horcasitas, A. S. and Snow, W. E., 1956, Geological resume of the Santa Eulalia district, Chihuahua, Mexico, p. 37-45, XXth International Geological Congress Guidebook for excursion A-2.

Hulin, Carlton D., 1929, Geology and mineralization at Pachuca, Mexico, p. 171-173, vol. 40, **Bulletin of the Geological Society of America.**

Jones, Warren, 1962, New fire agate find at Tula Hill, p. 113, April, **Lapidary Journal.**

Jones, Warren, 1962, The Guadalcazar-Realejo collecting area, p. 282-283, May, **Lapidary, Journal.**

Kasper, J., 1959, Stone and gem cutting in Mexico, p. 132-137, vol. 28, **The Gemmologist.**

Kunz, George Frederick, 1902, Gems and precious stones of Mexico, p. 55-93, vol. 32, **Transactions of the American Institute of Mining Engineers.**

Leechman, Frank, 1961, **The opal book,** 255 p., Australia, Ure Smith.

Lowther, G. K. and Marlow, G. C., 1956, Geology of the Parral area, p. 64-75, XXth International Geological Congress guidebook for excursion A-2.

Lyons, William J. and Young, Jack R., 1961, The colorful agates of northern Mexico, p. 258-259, 264, June, Lapidary Journal. Reprinted in **The agates of North America,** p. 78-79.

McAllister, James F. and Hernandoz, David, 1945, Quicksilver-antimony deposits of Huitzuco, Guerrero, Mexico, p. 49-71, **U.S. Geological Survey Bulletin 946-B.**

McConnell, Duncan, 1933, Garnets from Sierra Tlayacac, Morelos, Mexico, p. 25-29, vol. 18, **American Mineralogist.**

Mandarino, Joseph A. and Williams, Scott J., 1961, Five new minerals from Moctezuma, Sonora, Mexico, p. 2017, vol. 133, No. 3469, 23 June 1961, Washington, D.C., **Science.**

Merrill, F. J. H., 1906, The mining camps of Sinaloa, Mexico, p. 635-636, vol. 82, **Engineering and Mining Journal.**

Mrose, Mary E., 1948, Adamite from the Ojuela mine, Mapimi, Mexico; with notes on the occurrence by Dan E. Mayers and Francis A. Wise, vol. 33, p. 449-457, **American Mineralogist.**

Nininger, Robert D., 1954, **Minerals for atomic energy,** 367 p., D. Van Nostrand Company, Inc.

Ordonez, Ezequiel, 1902, The mining district of Pachuca, Mexico, p. 224-241, vol. 32, **Transactions of the American Institute of Mining Engineers.**

Ortiz-Asiain, Raul, 1956, Notes on Cerro de Mercado, p. 104-109, XXth International Geological Congress guidebook for excursion A-2.

Parliman, C. R., 1960, New agate fields in old Mexico, p. 4-7, April, **Lapidary Journal.**

Pettmecky, Ben J., 1960, Rock hunting in Mexico, p. 42-47, April, **Lapidary Journal.**

Prescott, Basil, 1916, The main mineral zone of the Santa Eulalia district, Chihuahua, p. 57-99, vol. 51, **Transactions of the American Institute of Mining Engineers.**

Rice, Helen M., 1959, A rockhound in Mexico, p. 22-25, 95, May, **Gems & Minerals.**

Rickard, T. A., 1924, The Ahumada lead mine and the ore deposits of the Los Lamentos Range, in Mexico, p. 365-373, vol. 118, **Engineering and Mining Journal-Press.**

Santillan, Manuel, 1936, **Carta geologico-minera del Estado de Durango,** p. 93-96, Instituto Geologico de Mexico.

Schmitt, Harrison, 1931, Geology of the Parral area of the Parral district, Chihuahua, Mexico, p. 268-290, general volume, **Transactions of the American Institute of Mining and Metallurgical Engineers.**

Segerstrom, Kenneth, 1961, Geology of the Bernal-Jalpan area, Estado de Queretaro, Mexico, p. 19-86, **U.S. Geological Survey Bulletin 1104-B.**

Simons, Frank S., and Mapes V., Eduardo, 1956, Geology and ore deposits of the Zimapan mining district, State of Hidalgo, Mexico, 128 p., **U.S. Geological Survey Professional Paper 284.**

Sinkankas, John, 1959, **Gemstones of North America,** 675 p., D. Van Nostrand Company, Inc.

Sinkankas, John, 1962, So-called psilomelane is chalcedony, p. 667, February, **Lapidary Journal.**

Sinkankas, John, 1962, Strawberry quartz . . . what is it?, p. 677-678, February, **Lapidary Journal.**

Stewart, Wendell O., 1940, The selenite caves of Naica, Mexico, p. 135-136, 193-195, vol. 8, No. 4. April, **The Mineralogist.**

Stone, J. B., 1956, Notes on the Sombrerete mining district of Zacatecas, p. 110, XXth International Geological Congress guidebook for excursion A-2.

Stone, J. B., 1956. Notes on the Fresnillo mining district of Zacatecas, p. 111-113, XXth International Geological Congress guidebook for excursion A-2.

Sutherland. Mason, 1955, Sonora is jumping, p. 215-246, vol. 107, No. 2, February, **National Geographic Magazine.**

Switzer, George; Foshag, W. F.; Murata, K. J.; and Fahey, J. J.; 1953, Re-examination of mosesite, p. 1225-1234, vol. 38, **American Mineralogist.**

Terry, T. Philip, 1935 and 1942, **Terry's guide to Mexico.**

Toothaker, Charles R., 1941, Calcite crystals with rhomboid tubes from Guanajuato, Mexico, p. 733-735, vol. 26, **American Mineralogist.**

van Geldern, J., 1960. Mineral and mining guide to western Mexico, 7 p., 1 map, published by the author, Box 24232, Los Angeles 24, California, $2.00.

Vaupell, C. W., 1937, The Huitzuco mercury mines, Guerrero, Mexico, (abstract), p. 196, vol. 32, **Economic Geology.**

Vaupell, C. W., 1941, Mercury deposits of Huitzuco, Guerrero, Mexico, p. 300-313, vol. 144, **Transactions of the American Institute of Mining and Metallurgical Engineers.**

Wandke. Alfred and Martinez, Juan. 1928, The Guanajuato mining district. Guanajuato, Mexico, p. 1-44, vol. 23, **Economic Geology.**

Wentworth, Irving H., 1913, The San Nicolas mining-district, San Nicolas, Tamaulipas, Mexico, p. 304-313, vol. 43, **Transactions of the American Institute of Mining Engineers.**

Wilson, Ivan F., 1956, The Naica mining district, Chihuahua, Mexico, p. 46-61, XXth International Geological Congress guidebook for excursion A-2.

Winchell, Horace V., 1921, Geology of Pachuca and El Oro, Mexico, p. 27-40, vol. 66, **Transactions of the American Institute of Mining and Metallurgical Engineers.**

GLOSSARY OF GEM, MINERAL AND MINING TERMS

Recently the pattern in the Spanish spelling of mineral names seems to be to follow the English and change the final e to a. In general from English to Spanish the following changes are usually made: z to c, ph to f, s to es, e to a, and y to i. Words with double letters in English sometimes drop one when translated into Spanish. There is always the exception and these general rules are not followd in every case. For the translation of common words consult a standard dictionary.

ENGLISH — SPANISH

A

acanthite — acantita
acid — ácido
acmite — acmita
actinolite — actinolita, -ota
adamite — adamita
adit — socavón
adularia — adularia
aerial — aéreo
agate — ágata
aguilarite — aguilarita
air — aire
alabandite — alabandita
alabaster — alabastro
albite — albita
allanite — allanita
allemontite — alemontita
allophane — alofanita
almandite — almandina
altaite — altaita
alumina — alúmina
aluminum — aluminio
alunite — alunita
alunogen — alunógena
amber — ámbar
amblygonite — ambligonita
amethyst — amatista
ammonia — amoníaco
amorphous — amorfo
amphibole var. asbestos — anfibola var. asbesto
analcime — analcima
analcite — analcita
analyze (to) — analizar
andalusite — andalusita, andalucita
andalusite var. chiastolite — andalusita var. quiastolita
andesine — andesina
andradite — andradita
anglesite — anglesita
anhydrite — anhidrita
ankerite — ankerita, anquerita
anorthite — anortita
anthophyllite — antofilita
anthracite coal — antracita
antimony — antimonio, estibio
anvil — bigornia, yunque
apatite — apatita
apophyllite — apofilita
aquamarine — aguamarina
aqueduct — acueducto
aragonite — aragonita
arc welding — soldadura electrica
arfvedsonite — arfvedsonita
argentiferous — argentífero
argentite — argentita, plata gris
arsenic — arsénico
arsenopyrite — arsenopirita, mispíquel
asbestos — asbesto, amianto
assay (to) — ensayer
assayer — ensayador
astrakanite (bloedite) — astrakanita
atacamite — atacamita
augite — augita
aurichalcite — auricalcita, auricalquita
auriferous — aurífero
autunite — autunita
axe — hacha
axinite — axinita
azurite — azurita

B

bar — barra, plancha
barite(s) — barita, baryta, barytes, baritina, baritita
barium — bario
barrel — barril
"bauxite" — bauxita
bentonite — bentonita
bentonite clays — arcillas bentoniticas
berthierine — bertierina
berthierite — bertierita
bertrandite — bertrandita
beryl — berilo
beryllium — berilio
beryllonite — berilonita
berzelianite — berzelianita
berzeliite — berzeliita
beudantite — beudantita
biereite — biereita
bindheimite — bindheimita
biotite — biotita
bismite — bismita
bismuth, native — bismuto nativo
bismuthinite — bismutinita
bismutite — bismutita
bituminous coal — carbón bituminoso
black — negro, - ra
blasting cap — cápsula, casquillo, fulminante
blister — burbuja
blister-copper — cobre roseta
blowpipe — soplete
board — tabla
boleite — boleíta
boracite — boracita
borax — bórax
bornite — bornita
boron — boro
botryoidal — arriñonada
bottom — fondo, suelo
boulangerite — boulangerita, bulangerita
bournonite — bournonita
box — caja
brass — latón
brazilianite — brasilianita
breccia — brecha
bridge — puente
brochantite — brochantita
bromine — bromo
bromyrite — bromirita
brucite — brucita
bucket — cubeta
burkeite — burkeita
bustamite — bustamita

C

cabazite — cabazita
cadmium — cadmio
calaverite — calaverita
calcareous — calcareo
calcareous tufa — toba
calcite — calcita
calcite var. manganocalcite — calcita var. manganocalcita
calcium — calcio
calomel — calomel
Cambrian — Cámbrico, -iano
candle — vela
canyon — barranca
carat — quilate
carbide — carburo
carbonate — carbonato
carboniferous — carbonífero
carminite — carminita
carnallite — carnalita
carnelian — carniola, cornalina
carnotite — carnotita
carphosiderite — carfosiderita
case — caja
cassiterite — casiterita
cauliflower — coliflor
cave — gruta, cueva
ceiling — cielo
celadonite — celadonita
celestite — celestina, celestita
cement — cemento, cimento
Cenozoic — Cenozoico
cerargyrite — cerargirita
cerium — cerio
cerussite — cerusita
cervantite — cervantita
cesium — cesio
chabazite — chabazita
chalcanthite — calcantita
chalcedony — calcedonia
chalcocite — calcocita, calcosina, chalcocita
chalcophanite — calcofanita
chalcopyrite — calcopirita, coribronce
chalcostibite — calcoestibita
chalk — creta
chamber — salón
charcoal — carbón
chemist — químico
chert — piedra córnea
chiastolite — quiastolita (andalusita)
chimney — chimenea
chlorine — cloro
chlorite — clorita
chondrodite — condrodita
chromite — cromita
chromium — cromo
chrysoberyl — crisoberilo
chrysocolla — crisócola
chrysolite — crisólito, peridoto
chrysotile — crisotilo
cinnabar — cinabrio
clay — arcilla, argila, argilla, barro
cleavelandite — clevelandita
cliff — bufa
coal — carbón, carbón de piedra, carbón mineral
cobalt — cobalto
cobaltite — cobaltita
cobaltocalcite — cobaltocalcita
cocinerite — cocinerita
cold — frío, - ía
colemanite — colemanita
collect (to) — recoger
collector — colector
coloring earth — tierras colorantes
columbite — columbita
columbium — columbio
common salt — sal común
conchoidal — concoide
concretion — concreción
conglomerate — conglomerado
contact — contacto
contractor — contratista
copiapite — copiapita
copper — cobre
copper, native — cobre virgen, cobre nativo
coral — coral
cord — cuerda
cordierite — cordierita
coronadite — coronadita
correct — direcho
corundum — corundo, corindón
cosalite — cosalita
country rock — montaña
covellite — covellina, covellita, covelita
cover — tapa
Cretaceous — Cretáceo, Cretácico, Cretacio
crevice — grieta
cristobalite — cristobalita
crocidolite — crocidolita
crocoite — crocoíta
crocosit — crocosita
cross — cruz
crowbar — barreta
crush (to) — machacar
crusher — machucador, triturador
crust — corteza
cryolite — criolita
cryptomelane — criptomelán
crystal — cristal
crystalline — cristalina, cristalino
cubic — cúbico
cumengite — cumengeíta
cuprite — cuprita
cuprouranite — cuprouranita
curve — curva
cyanide — cianuro

D

danburite — danburita
danger — peligro
datolite — datolita
daubreelite — daubreelita
density — densidad
descloizite — descloizita
desert — desierto
Devonian — Devoniano, -ico
diabase — diabasa
diamond — diamante
diaspore — diásporo
diatomaceous — diatomeas
diatomaceous earth — tierras diatomeas, tierras diatomáceas
diatomite — diatomita
dike — dique
diopside — diópsida, diópsido
dioptase — dioptasa
diorite — diorita
dip — recuesto, echado
duftite — duftita
discovery — hallazgo
district — zona
dolerite — dolerita
dolomite — dolomita
dome — doma
domeykite — domeikita
drag mill — arrastre, a circular trough in which drags are pulled around for grinding and amalgamating ores.
drill — taladro
drill bit — broca
drill hole — barreno
drill rod — taladro
duftite — duftita
dumontite — dumontita

dumortierite — dumortierita
durangite — duranguita
dynamite — dinamita

E

east — este, oriente
eat (to) — comer
effervescence — efervescencia
embankment — terraplén
embolite — embolita
emerald — esmeralda
emery — esmeril
empty — vacío, - ía
enargite — enargita
engineer — ingeniero (Ing.)
enstatite — enstatita
Eocene — Eoceno
epidote — epidota, epidoto
epsomite — epsomita
erythrite — eritrina, eritrita
excavate (to) — excavar
exploitation — disfrute

F

facet of crystal — cara
fayalite — fayalita
feldspar — feldespato
feldspar sands — arenas feldespaticas
felsite — felsita
fergusonite — fergusonita
ferricyanide — ferricanuro
ferrocyanide — ferrocianuro
ferruginous — ferroso
fibrous — fibrosa
file — lima
fissure — fisura
fissure vein — veta
flint — pedernal, piedra córnea
flame — llama
fluorescent — fluorescente
fluorescence — fluorescencia
fluorite — fluorita
fold — pliegne
formation — formación
fossil — fósil
fossiliferous — fosilifera
fracture — fractura
franklinite — franklinita, franclinita
fuller's earth — tierras de fuller
fuse — cañuela, mecha

G

gabbro — gabbro
galena — galena
gangue — guija, ganga
garnet — granate
garnet var. almandite — granate var. almandina, almandita
garnet var. grossularite — granate var. grosularita
garnet var. pyrope — granate var. piropo
garnierite — garnierita
gas — gaz
gearksutite — gearksutita
gem — gema, piedra preciosa
geode — geoda
geologist — geólogo
geology — geología
germanium — germanio
gersdorffite — gersdorffita
geyserite (siliceous sinter) — geiserita
gibbsite — gibsita
glaserite — glaserita
glass — vidrio
glauberite — gláuberita
glaucodot — glaucodoto
globular — globoso, - sa
goethite — goetita
gold — oro
gold-bearing — aurífero
gold, native — oro nativo
gorge — barranca
granite — granito
granitic — granítico
granular — granular
graphite — grafito, grafita, lápiz
greenockite — greenockita
grind (to) — moler
grossularite — groslarita, grosularia
ground — terreno
guadalcazarite — guadalcazarita
guanajuatite — guanajuatita
gulf — golfo
gypsum — yeso
gypsum var. alabaster — yeso var. alabastro
gypsum var. selenite — yeso var. selenita

H

halite — halita
halloysite — halloysita
hammer — martillo
hand-hammer — martillo
hand specimen — piedra de mano
hanksite — hanksita
hardness — dureza
harmotome — harmotomia
hatchet — hachita
head — cabeza
hedenbergite — hedenbergita, hedembergita
helper — peón
hematite — hematita
hematite var. specular — hematita var. hierro espejado
hemimorphite — hemimorfita, calamina, calamita
hessite — hessita
heulandite — heulandita
hexagonal — exagonal
hill — cerro, colina
hisingerite — hisingerita
hoist — malacate
hollandite — holandita
hornblende — hornblenda
hot — caliente
huebnerite — huebnerita
hyalite — hialita
hydrogen — hidrógeno
hydrozincite — hidrozincita
hypersthene — hiperstena

I

ice — hielo
idocrase — idocrasa, vesuviana
igneous — ígneo, vulcanio
ilmenite — ilmenita
ilvaite — ilvaíta
inclined — oblicuo
inclusion — partícula extraña, intrusión
Indian — indio
ingot — barra, plancha, lingote
intrusive rock — roca intrusiva
iodyrite — iodirita
iron — fierro, hierro
iron meteorite — fierro meteórico

J

jade — jade, nefrita, piedra nefrítica, piedra de ijada
jadeite — jadeita
jalpaite — jalpaita

jamesonite — jamesonita
jarosite — jarosita
jasper — jaspe
jet — azabache
Jurassic — Jurásico

K

kaolin — kaolín, caolín
kaolinite — kaolinita
kermesite — antimonio rojo
kernite — quernita
knebelite — knebelita
knife, large — machete

L

labradorite — labradorita
lapis lazuli — lapislázuli, lazulita
laterites — lateritas
laumontite — laumontita
law — ley
lazulite — lazulita
lead — plomo
lens — lente
lepidolite — lepidolita
leucite — leucita
level — nivel
lever — palanca
lightweight aggregates — agregados ligeros
lignite — lignito
lime — cal
limestone — caliza, piedra de cal
"limonite" — limonita
lithium — litio
lithology — litología
livingstonite — livingstonita
loellingite — loelingita
longitude — longitud
lubricant — lubricante
lumber — madera
luster — brillo

M

magnesite — magnesita
magnesium — magnesio
magnet — iman
magnetic — magnético
magnetite — magnetita
malachite — malaquita, cobre verde
manganese — manganeso
manganite — manganita
marble — mármol
marcasite — marcasita, marquesita
margarite — margarita
marl — marga
marmatite — marmatita
martite — martita
mass — masa
massicot — masicote
massive — masiva
melanterite — melanterita
mercury — mercurio
Mesozoic — Mesozoico
metacinnabar — metacinabrio
metalliferous mineral — metal
metallurgy — metalurgia
metamorphic — metamórfico
metamorphic rocks — rocas metamórficas
metamorphism — metamorfosis
miargyrite — miargirita
mica — mica
microcline — microclina
mill — molino
millerite — millerita
mimetite — mimetena, mimetita
mine — mina
mine entrance — bartolina
mine owner — minero
miner — minero
miner, gopher — gambusino
mineral — mineral
miner's cap — boina
mine shaft — tiro, pozo, apique
mining — minero, minería, minera
mining claim — propiedad minera, pertenencia
mining district — comarca minera
mining tunnel — socavón
minium — minio
Miocene — Mioceno
mirabilite — mirabilita
Mississippian — Carbonifero Inferior, Misisipiano
moldavite — moldavita
molybdenite — molíbdenita
molybdenum — molíbdeno
monazite — monacita
montmorillonite — montmorillonita
mould — molde
mountain range (a) — sierra
mouth of mine or tunnel — boca
mud — lodo
mule — mula
muscovite — muscovita, moscovita
museum — museo

N

nail — clavo
nagyagite — nagyagita
native — nativo
natrolite — natrolita
natron — natrón
naumannite — naumanita
nepheline — nefelina
nephrite — nefrita
niccolite — nicolita, niquelina
nickel — níquel
niobium — niobio
niter — nitre, nitro (salitre)
nitrogen — nitrógeno
nodules — nódulos
north — norte
nugget — pepita

O

oak — encina
obsidian — obsidiana
ocher — ocre
octahedron — octaedro
office — oficina
Oligocene — Oligoceno
oligoclase — oligoclasa
olivine — olivino, peridoto
onofrite — onofrita
onyx — ónice, ónix
opal — ópalo
opal, common — ópalo var. común
opalized wood — madera opalino
opal var. hyalite — ópalo var. hialita
optical — optica
ordoñezite — ordoñezita
Ordovician — Ordoviciano
ore deposit — vacimiento
ore shoot — chorrera
ores — frutos
orpiment — oropimente
orthoclase — ortoclasa, ortosa
owner — dueño
oxide — óxido
oxygen — oxígeno

P

Paleocene — Paleoceno
Paleozoic — Paleozoico

palladium — paladio
parallel — paralelo
pearl — perla
peak — picacho
peat — turba
pebble — guijarro
pectolite — pectolita
peganite — peganita
pegmatite — pegmatita
pegmatites — pegmatíticos
pentlandite — pentlandita
Pennsylvanian — Pensilvano
peridot — peridoto, crisólito, olivino
perlite — perlita
Permian — Pérmico, -iano
petalite — petalita
petrified wood — madera petrificada
petzite — petzita
phacolite — facolita
phenacite — fenacita
phenakite — fenacita
phillipsite — filipsita
phlogopite — flogopita
phonolite — fonolito
phosphates — fosfatos
phosphorescence — fosforescencia
phosphorite — fosforita
phosphorus — fósforo
pickaxe — pico
picotite — picotita
pig — plancha
pipe — tubo
pit — cata
pitchblende — pechblenda
placer — placer, lavadero
plank — tabla
plate — plancha
platinum — platino
platinum ore — platina
Pleistocene — Pleistoceno
Pliocene — Plioceno
plumbojarosite — plumbojarosita
pocket — bolsa
pocket of ore — bolsón
polybasite — polibasita, teposteles
port — puerto
pottery — loza
prismatic — prismatico
profile — perfil
prospect (to) — catar, catear
prospect hole — cata
prospector — gambusino, buscador, cateador
prospectors — buscónes
proustite — proustita
pseudoboleite — seudoboleíta
psilomelane — psilomelán, psilomelano
pumice — pumita, piedra pómez, pumicita, pomez
pump — bomba
pyramidal — piramidal
pyrargyrite — pirargirita, plata roja
pyrite — pirita
pyrolusite — pirolusita
pyromorphite — piromorfita
pyrophyllite — pirofilita
pyrrhotite — pirrotina, pirrotita

Q

quartz — cuarzo
quartz var. agate — cuarzo var. ágata
quartz var. amethyst — cuarzo var. amatista
quartz var. chalcedony — cuarzo var. calcedonia
quartz var. rose — cuarzo var. rosa, cuarzo rosa
quartzite — cuarcita
Quaternary — Cuaternario
quicksilver — azoque

R

railroad — ferrocarril
rake — rastrillo
rare earths — tierras raras
ravine — barranca
realgar — rejalgar
Recent — Reciente
red — colorado
refractory clays — barros refractarios, arcillas refractarias
regular — formal
rhodochrosite — rodocrosita, rodocrusita
rhodonite — rhodonita, rodonita
rhyolite — riolita
rich — rico
river — río
rivet — remache
rock — roca, peña
roof — cielo
rope, light — reata
rose quartz — cuarzo rosa
ruby — rubí
ruby silver — cardenillo
ruby spinel — espinela, balaj
rutile — rutilo

S

salt — sal
salt deposit — salina
salted — salada, salado
salt mine — salinas
saltpeter — nitro (salitre)
salts, nitrogenous — sales nitrogenadas
saltworks — salinas
salty — salada, salado
sample — muestra
sand — arena
sandstone — arenisca, piedra arenisca
sapphire — zafir, zafiro
sartorite — sartorita
scapolite — escapolita
scheelite — scheelita
schist — esquisto
screen — criba
screwdriver — destornillador
seam — criadero
sedimentary — sedimentario
selenite — selenita
selenium, native — selenio
sepiolite — sepiolita
sericite — sericita
serpentine — serpentina
shaft (mine) — pozo, tiro, apique
shale — esquisto, pizarra
shovel — pala
siderite — siderita
silica — sílice
silicate — silicato
siliceous — silícea, silíceo
silicon — silicio
sillimanite — silimanita
Silurian — Siluriano, -ico
silver — plata
silver, native — plata nativa, plata blanca
skutterudite — skutterudita
sledge hammer — marro
slickensides — salbanda
smaltite — esmaltina, esmaltita
smelt (to) — fundir
smelter — fundidor
smithsonite — smithsonita
sodalite — sodalita

sodium — sodio
sodium salts — sales sodicas
south — sud, sur
specimens — muestras, especímenes, riscos
specularite — especularita, hematite brillante, hierro especular
spessartite — espesartita
sphalerite — esfalerita, blenda
sphene — esfena, esfeno, titanita
spinel, ruby — espinela, balaj
spodumene — espodumena
spoon — cuchara
stalactites — estalactitas
stalagmites — estalagmitas
stannite — estannita
staurolite — estaurolita
steam — vapor
steatite — esteatita
stephanite — estefanita
stibiconite — estibiconita
stibnite — estibina, estibnita
stick — palo
stilbite — estilbita
stone — piedra
strata — estratos
strike — dirección
stromeyerite — stromeyerita
strontianite — estroncianita
strontium — estroncio
sulfide of copper — chalcosina
sulfide ores — azufrones
sulfides — sulfuros
sulfur — azufre
summit — cima, cumbre
superintendent — encargado
surface — superficie
syenite — sienita
sylvanite — silvanita

T

tailings — colas
talc — talco
talc var. steatite — talco var. esteatita
tank — tanque
tantalite — tantalita
teepleite — teeplita
telluride — telururo
tellurite — telurita
tellurium — teluro
tenorite var. melaconite — tenorita var. melaconita
Tertiary — Terciario
tetrahedrite — tetraedrita, cobre negro
texture — textura
thermonatrite — termonatrita
thomsonite — thomsonita
thorium — torio
thulite — tulita
timber — madera
tin — estaño
titanite — titanita
titanium — titanio
ton — tonelada
tongs — tenazas
tooth — diente
top — cumbre
topaz — topacio
torbernite — torbernita
to the right — direcha
tourmaline — turmalina
trachyte — traquito
tramway — tranvía
translucent — translúcida
transparent — transparente
travertine — travertino, toba
tremolite — tremolita
Triassic — Triásico
tridymite — tridimita
troilite — troilita
trona — trona
tube — tubo
tuff, volcanic — toba
tungsten — tungsteno
turgite — turgita
turquois — turquesa

U

United States of America — EE.UU.
uralite — uralita
uraninite — uraninita
uranium — uranio, urano

V

valentinite — antimonio blanco
valley — valle
valve — válvula
vanadinite — vanadinita
vanadium — vanadio
variscite — variscita, varisita
vein — veta
vein formation — panino
vein matter — borra
velocity — velocidad
ventilation — ventilación
vermiculite — vermiculita
vesuvianite — vesuviana, idocrasa, vesuvianita
vitreous — vítreo
volcano — volcán
vug — soyote

W

wad — wad
wall rock — peña
waste rock — escombros, rezaga
watchman — velador
water — agua
wedge — cuña
weight — peso
west — oeste
white — blanca, blanco
willemite — willemita
witherite — witerita
wolframite — wolframita, volframita
wollastonite — wollastonita
wulfenite — wulfenita
wurtzite — wurtzita

Z

zinc — cinc, zinc
zincblende (sphalerite) — blenda, copalillo
zincite — cincita
zircon — circón
zirconium — zirconio, circonio
zoisite — zoicita
zone — zona

SPANISH — ENGLISH

A

acantita — acanthite
ácido — acid
acmita — acmite
actinolita — actinolite
actinota — actinolite
acueducto — aqueduct
adamita — adamite
adularia — adularia
aéreo — aerial
ágata — agate
agregados ligeros — lightweight aggregates
agua — water
aguamarina — aquamarine
aguilarita — aguilarite
aire — air
alabandita — alabandite
alabastro — alabaster
albita — albite
alemontita — allemontite
almandina — almandite
alofanita — allophane
altaita — altaite
alúmina — alumina
aluminio — aluminum
alunita — alunite
alunógena — alunogen
allanita — allanite
amatista — amethyst
ámbar — amber
ambligonita — amblygonite
amianto — asbestos
amoníaco — ammonia
amorfo — amorphous
analcima — analcime
analcita — analcite
analizar — to analyze
andalucita — andalusite
andalusita — andalusite
andalusita var. quiastolita — andalusite var. chiastolite
andesina — andesine
andradita — andradite
anfibola var. asbesto — amphibole var. asbestos
anglesita — anglesite
anhidrita — anhydrite
ankerita — ankerite
anortita — anorthite
anquerita — ankerite
antimonio — antimony
antimonio blanco — valentinite
antimonio rojo — kermesite
antofilita — anthophyllite
antracita — anthracite coal
apatita — apatite
apique — mine shaft
apofilita — apophyllite
aragonita — aragonite
arcilla — clay
arcillas bentoniticas — bentonite clays
arcillas refractarias — refractory clays
arena — sand
arenas feldespaticas — feldspar sands
arenisca — sandstone
arfvedsonita — arfvedsonite
argentífero — argentiferous
argentita — argentite
argila — clay
argilla — clay
arsénico — arsenic
arsenopirita — arsenopyrite
arrastre — drag mill
arriñonada — botryoidal
asbesto — asbestos
astrakanita — astrakanite (bloedite)
atacamita — atacamite
augita — augite
auricalcita — aurichalcite
auricalquita — aurichalcite
aurífero — auriferous (gold-bearing)
autunita — autunite
axinita — axinite
azabache — jet
azoque — quicksilver
azufre — sulfur
azufrones — sulfide ores
azules — blue opals
azurita — azurite

B

balaj — ruby spinel
bario — **barium**
barita — **barite**
baritina — **barite**
baritita — barite
bartolina — mine entrance
baryta — barite
barytes — barite
barra — bar, ingot
barranca — canyon, gorge, ravine
barreno — drill hole
barreta — crowbar
barril — barrel
barro — clay
barros refractarios — refractory clays
bauxita — "bauxite"
bentonita — bentonite
berilio — beryllium
berilo — beryl
berilonita — beryllonite
bertierina — berthierine
bertierita — berthierite
bertrandita — bertrandite
berzelianita — berzelianite
berzeliita — berzeliite
beudantita — beudantite
biereita — biereite
bigornia — anvil
bindheimita — bindheimite
biotita — biotite
bismita — bismite
bismutinita — bismuthinite
bismutita — bismutite
bismuto — bismuth
bismuto nativo — native bismuth
blanca — white
blanco — white
blenda — sphalerite, zincblende
boca — mouth of mine or tunnel
boina — miner's cap
boleíta — boleite
bolsa — pocket
bolsón — pocket of ore
bomba — pump
boracita — boracite
bórax — borax
bornita — bornite
boro — boron
borra — vein matter
boulangerita — boulangerite
bournonita — bournonite
brasilianita — brazilianite
brecha — breccia
brillo — luster
broca — drill bit
brochantita — brochantite
bromirita — bromyrite
bromo — bromine
brucita — brucite
bufa — cliff
bulangerita — boulangerite
burbuja — blister

burkeita — burkeite
buscador — prospector
buscónes — prospectors
bustamita — bustamite

C

cabazita — cabazite
cabeza — head
cadmio — cadmium
caja — box, case
cal — lime
calamina — hemimorphite
calamita — hemimorphite
calaverita — calaverite
calcantita — chalcanthite
calcareo — calcareous
calcedonia — chalcedony
calcio — calcium
calcita — calcite
calcita var. manganocalcita — calcite var. manganocalcite
calcocita — chalcocite
calcoestibita — chalcostibite
calcofanita — chalcophanite
calcopirita — chalcopyrite
calcosina — chalcocite
caliente — hot
caliza — limestone
calomel — calomel
Cambriano — Cambrian
Cámbrico — Cambrian
cañuela — fuse
caolín — kaolin
cápsula — blasting cap
cara — facet of a crystal
carbón — coal, charcoal
carbonato — carbonate
carbón bituminoso — bituminous coal
carbón de piedra — coal
carbonífero — carboniferous
Carbonifero Inferior — Mississipian
carbón mineral — coal
carburo — carbide
cardenillo — ruby silver
carfosiderita — carphosiderite
carminita — carminite
carnalita — carnallite
carniola — carnelian
carnotita — carnotite
casiterita — cassiterite
casquillo — blasting cap
cata — pit, prospect hole
catar — to prospect
cateador — prospector
catear — to prospect
celadonita — celadonite
celestina — celestite
celestita — celestite
cemento — cement
Cenozoico — Cenozoic
cerargirita — cerargyrite
cerio — cerium
cerusita — cerussite
cervantita — cervantite
cerro — hill
cesio — cesium
cianuro — cyanide
cielo — roof, ceiling
cima — summit
cimento — cement
cinabrio — cinnabar
cinabrio negro — black cinnabar
cinc — zinc
cincita — zincite
circón — zircon
circonio — zirconium
clavo — nail
clevelandita — cleavelandite
clorita — chlorite
cloro — chlorine
cobaltita — cobaltite
cobalto — cobalt
cobaltocalcita — cobaltocalcite
cobre — copper
cobre nativo — native copper
cobre negro — tetrahedrite
cobre roseta — blister copper
cobre verde — malachite
cobre virgen — native copper
cocinerita — cocinerite
colas — tailings
colector — collector
colemanita — colemanite
coliflor — cauliflower
colina — hill
colorado – red
columbio — columbium
columbita — columbite
comarca minera — mining district
comer — to eat
concoide — conchoidal
concreción — concretion
condrodita — chondrodite
conglomerado — conglomerate
contacto — contact
contratista — contractor
copalillo — zincblende (sphalerite)
copiapita — copiapite
coral — coral
cordierita — cordierite
coribronce — chalcopyrite
corindón — corundum
cornalina — carnelian
coronadita — coronadite
corteza — crust
corundo — corundum
cosalita — cosalite
covelita — covellite
covellina — covellite
covellita — covellite
creta — chalk
Cretáceo, -cico, -cio — Cretaceous
Cretácico — Cretaceous
Cretáceo — Cretaceous
criadero — seam
criba — screen
criolita — cryolite
criptomelán — cryptomelane
crisoberilo — chrysoberyl
crisócola — chrysocolla
crisólito — chrysolite, peridot
crisotilo — chrysotile
cristal — crystal
cristalina -no — crystalline
cristobalita — cristobalite
crocidolita — crocidolite
crocoíta — crocoite
crocosita — crocosite
cromita — chromite
cromo — chromium
cruz — cross
cuarcita — quartzite
cuarzo — quartz
cuarzo rosa — rose quartz
cuarzo var. ágata — quartz var. agate
cuarzo var. amatista — quartz var. amethyst
cuarzo var. calcedonia — quartz var. chalcedony
cuarzo var. rosa — quartz var. rose
Cuaternario — Quaternary
cubeta — bucket
cúbico — cubic
cuchara — spoon
cuerda — cord
cueva — cave
cumbre — top, summit
cumengeíta — cumengite
cuña — wedge
cuprita — cuprite
cuprouranita — cuprouranite

curva — curve

CH

chabazita — chabazite
chalcocita — chalcocite
chalcosina — sulfide of copper
chimenea — chimney
chorrera — ore shoot

D

danburita — danburite
datolita — datolite
daubreelita — daubreelite
densidad — density
descloizita — descloizite
desierto — desert
destornillador — screwdriver
Devoniano, -ico — Devonian
diabasa — diabase
diamante — diamond
diásporo — diaspore
diatomeas — diatomaceous
diatomita — diatomite
diente — tooth
dinamita — dynamite
diópido, -da — diopside
dioptasa — dioptase
diorita — diorite
dique — dike
dirección — strike
direcha — to the right
direcho — correct
disfrute — exploitation
dolerita — dolerite
dolomita — dolomite
doma — dome
domeikita — domeykite
dueño — owner
duftita — duftite
dumontita — dumontite
dumortierita — dumortierite
duranguita — durangite
dureza — hardness

E

EE.UU. — United States of America
echado — dip
efervescencia — effervescence
embolita — embolite
enargita — enargite
encargado — superintendent
encina — oak
ensayador — assayer
ensayer — to assay
enstatita — enstatite
Eoceno — Eocene
epidota — epidote
epidoto — epidote
epsomita — epsomite
eritrina — erythrite
eritrita — erythrite
escapolita — scapolite
escombros — waste rock
esfalerita — sphalerite
esfena, esfeno — sphene
esmaltina — smaltite
esmaltita — smaltite
esmeralda — emerald
esmeril — emery
espécimen — specimen
especularita — specularite
espesartita — spessartite
espinela — ruby spinel
espodumena — spodumene
esquisto — schist, shale
estalactitas — stalactites
estalagmitas — stalagmites
estannita — stannite
estaño — tin
estaurolita — staurolite
este — east
esteatita — steatite
estefanita — stephanite
estibiconita — stibiconite
estibina — stibnite
estibio — antimony
estibnita — stibnite
estilbita — stilbite
estratos — strata
estroncianita — strontianite
estroncio — strontium
exagonal — hexagonal
excavar — to excavate

F

facolita — phacolite
fayalita — fayalite
feldespato — feldspar
feldespatos — feldspars
felsita — felsite
fenacita — phenakite
fergusonita — fergusonite
ferricanuro — ferricyanide
ferrocarril — railroad
ferrocianuro — ferrocyanide
ferroso — ferruginous
fibrosa — fibrous
fierro — iron
fierro meteórico — iron meteorite
filipsita — phillipsite
fisura — fissure
flogopita — phlogopite
fluorescencia — fluorescence
fluorescente — fluorescent
fluorita — fluorite
fondo — bottom
fonolito — phonolite
formación — formation
formal — regular
fosfatos — phosphates
fosforescencia — phosphorescence
fosforita — phosphorite
fósforo — phosphorus
fósil — fossil
fosilifera — fossiliferous
fractura — fracture
franclinita — franklinite
franklinita — franklinite
frío, - ía — cold
frutos — ores
fulminante — blasting cap
fundidor — smelter
fundir — to smelt

G

gabbro — gabbro
galena — galena
gambusino — prospector, gopher miner, etc.
ganga — gangue
garnierita — garnierite
gaz — gas
gearksutita — gearksutite
geiserita — geyserite (siliceous sinter)
gema — gem
geoda — geode
geología — geology
geólogo — geologist
germanio — germanium
gersdorffita — gersdorffite
gibsita — gibbsite
glaserita — glaserite
gláuberita — glauberite
glaucodoto — glaucodot
globosa, -so — globular
goetita — goethite
golfo — gulf
grafita — graphite
grafito — graphite
granate — garnet
granate var. almandina, almandita — garnet var. almandite
granate var. grosularita — garnet var. grossularite

granate var. piropo — garnet var. pyrope
granítico — granitic
granito — granite
granular — granular
greenockita — greenockite
grieta — crevice
grosularia — grossularite
grosularita — grossularite
grutas — caves
guadalcazarita — guadalcazarite
guanajuatita — guanajuatite
guija — gangue
guijarro — pebble

H

hacha — axe
hachita — hatchet
halita — halite
hallazgo — discovery
halloysita — halloysite
hanksita — hanksite
harmotomia — harmotome
hedembergita — hedenbergite
hedenbergita — hedenbergite
hematita — hematite
hematita var. hierro espejado — hematite var. specular
hematites — hematite
hematites brillante — specularite
hemimorfita — hemimorphite
hessita — hessite
heulandita — heulandite
hialita — hyalite
hidrógeno — hydrogen
hidrozincita — hydrozincite
hielo — ice
hierro — iron
hierro especular — specularite
hiperstena — hypersthene
hisingerita — hisingerite
holandita — hollandite
hornblenda — hornblende
huebnerita — huebnerite

I

idocrasa — idocrase, vesuvianite
ígneo — igneous
ilmenita — ilmenite
ilvaíta — ilvaite
iman — magnet
indio — Indian
Ing. — engineer
ingeniero — engineer
intrusión — inclusion
iodirita — iodyrite

J

jade — jade
jadeita — jadeite
jalpaita — jalpaite
jamesonita — jamesonite
jarosita — jarosite
jaspe — jasper
Jurásico — Jurassic

K

kaolín — kaolin
kaolinita — kaolinite
knebelita — knebelite

L

labradorita — labradorite
lapislázuli — lapis lazuli
lápiz — graphite
lateritas — laterites
latón — brass
laumontita — laumontite
lavadero — placer
lazulita — lazulite, lapis lazuli
lente — lens
lepidolita — lepidolite
leucita — leucite
ley — law
lignito — lignite
lima — file
limonita — "limonite"
lingote — ingot
litio — lithium
litología — lithology
livingstonita — livingstonite
lodo — mud
loelingita — loellingite
longitud — longitude
loza — pottery
lubricante — lubricant

LL

llama — flame

M

machacar — to crush
machete — large knife
machucador — crusher
madera — timber, lumber
madera opalina — opalized wood
madera petrificada — petrified wood
magnesio — magnesium
magnesita — magnesite
magnético — magnetic
magnetita — magnetite
malacate — hoist
malaquita — malachite
manganeso — manganese
manganita — manganite
marcasita — marcasite
marga — marl
margarita — margarite
marmatita — marmatite
mármol — marble
marquesita — marcasite
martillo — hand hammer
martita — martite
marro — sledge hammer
masa — mass
masicote — massicot
masiva — massive
mecha — fuse
melanterita — melanterite
mercurio — mercury
Mesozoico — Mesozoic
metacinabrio — metacinnabar
metal — metalliferous mineral
metalurgia — metallurgy
metamórfico — metamorphic
metamorfosis — metamorphism
miargirita — miargyrite
mica — mica
microclina — microcline
millerita — millerite
mimetena — mimetite
mimetita — mimetite
mina — mine
minera — mining
mineral — mineral
minería — mining
minero — miner, mine owner, mining
minio — minium
Mioceno — Miocene
mirabilita — mirabilite
mispíquel — arsenopyrite
Misisipiano — Mississippian
moldavita — moldavite
molde — mould
moler — to grind
molíbdenita — molybdenite
molíbdeno — molybdenum
molino — mill
monacita — monazite
montaña — country rock
monte — mountain
montmorillonita — montmorillonite

moscovita — muscovite
muestra — sample, specimen
mula — mule
mun. — municipio, equivalent to a county in the United States
municipio (mun.) — the equivalent of a county in the United States
municipio — municipality
muscovita — muscovite
museo — museum

N

nagyagita — nagyagite
nativo — native
natrolita — natrolite
natrón — natron
naumanita — naumannite
nefelina — nepheline
nefrita — nephrite, jade
negra — black
nicolita — niccolite
niobio — niobium
níquel — nickel
niquelina — niccolite
nitre — niter
nitro (salitre) — niter, saltpeter
nitrógeno — nitrogen
nivel — level
nódulos — nodules
norte — north

O

oblicuo — inclined
obsidiana — obsidian
ocre — ocher
octaedro — octahedron
oeste — west
oficina — office
oligoclasa — oligoclase
Oligoceno — Oligocene
olivino — olivine, peridot
ónice — onyx
ónix — onyx
onofrita — onofrite
ópalo — opal
ópalo var. común — common opal
ópalo var. hialita — opal var. hyalite
optica — optical
ordoñezita — ordoñezite
Ordoviciano — Ordovician
oriente — east
oro — gold
oro nativo — native gold
oropimente — orpiment
ortoclasa — orthoclase
ortosa — orthoclase
óxido — oxide
oxígeno — oxygen

P

pala — shovel
paladio — palladium
palanca — lever
Paleoceno — Paleocene
Paleozoico — Paleozoic
palo — stick
panino — vein formation
paralelo — parallel
partícula extraña — inclusion
pechblenda — pitchblende
pectolita — pectolite
pedernal — flint
peganita — peganite
pegmatita — pegmatite
pegmatíticos — pegmatites
peligro — danger
peña — rock, wall rock
Pensilvano — Pennsylvanian
pentlandita — pentlandite
peón — helper
pepita — nugget
perfil — profile
peridoto — olivine, peridot, chrysolite
perla — pearl
perlita — perlite
Permiano — Permian
Pérmico — Permian
pertenencia — mining claim
peso — weight
petalita — petalite
petzita — petzite
picacho — peak
pico — pick-axe
picotita — picotite
piedra — stone
piedra arenisca — sandstone
piedra córnea — flint, chert
piedra de cal — limestone
piedra de ijada — jade
piedra de mano — hand specimen
piedra nefrítica — jade
piedra pómez — pumice
piedras preciosas — precious stones, gems
piramidal — pyramidal
pirargirita — pyrargyrite
pirita — pyrite
pirofilita — pyrophyllite
pirolusita — pyrolusite
piromorfita — pyromorphite
pirrotina — pyrrhotite
pirrotita — pyrrhotite
pizarra — shale
placer — placer
plancha — pig, ingot, bar, plate
plata — silver
plata blanca — native silver
plata gris — argentite
plata nativa — native silver
plata roja — pyrargyrite
platina — platinum ore
platino — platinum
Pleistoceno — Pleistocene
pliegne — fold
Plioceno — Pliocene
plomo — lead
plumbojarosita — plumbojarosite
polibasita — polybasite
pomez — pumice
pozo — mine shaft
prismatico — prismatic
propiedad minera — mining claim
proustita — proustite
psilomelán — psilomelane
psilomelano — psilomelane
puente — bridge
puerto — port
pumicita — pumice
pumita — pumice

Q

quernita — kernite
quiastolita (andalusita) — chiastolite
quilate — carat
químico — chemist

R

rastrillo — rake
reata — light rope
Reciente — Recent
recoger — to collect
recuesto — dip
rejalgar — realgar
remache — rivet
rezaga — waste rock
rhodonita — rhodonite
rico — rich
río — river
riolita — rhyolite
riscos — specimens

roca — rock
roca intrusiva — intrusive rock
roca metamórfica — metamorphic rock
rodocrosita — rhodochrosite
rodocrusita — rhodochrosite
rodonita — rhodonite
rubí — ruby
rutilo — rutile

S

sal — salt
salado, -da — salty, salted
salbanda — slickensides
sal común — common salt
sales nitrogenadas — nitrogenous salts
sales sodicas — sodium salts
salina — salt deposit
salinas — salt mine, saltworks
salón — chamber
sartorita — sartorite
scheelita — scheelite
sedimentario — sedimentary
selenio — selenium, native selenium
selenita — selenite
sepiolita — sepiolite
sericita — sericite
serpentina — serpentine
seudoboleíta — pseudoboleite
siderita — siderite
sienita — syenite
sierra — mountain range
silicato — silicate
sílice — silica
silícea — siliceous
silíceo — siliceous
silicio — silicon
silimanita — sillimanite
Siluriano, -ico — Silurian
silvanita — sylvanite
skutterudita — skutterudite
smithsonita — smithsonite
socavón — mining tunnel, adit
sodalita — sodalite
sodio — sodium
soldadura electrica — arc welding
soplete — blowpipe
soyote — vug
stromeyerita — stromeyerite
sud — south
suelo — bottom
sulfuros — sulfides
superficie — surface
sur — south

T

tabla — board, plank
taladro — drill, drill rod
talco — talc
talco var. esteatita — talc var. steatite
tanque — tank
tantalita — tantalite
tapa — cover
teeplita — teepleite
telurita — tellurite
teluro — tellurium
telururo — telluride
tenazas — tongs
tenorita var. melaconita — tenorite var. melaconite
teposteles — polybasite
Terciario — Tertiary
termonatrita — thermonatrite
terraplén — embankment
terreno — ground
tetraedrita — tetrahedrite
textura — texture
thomsonita — thomsonite
tierras colorantes — coloring earths
tierras de füller — fuller's earth
tierras diatomáceas — diatomaceous earth
tierras diatomeas — diatomaceous earth
tierras raras — rare earths
tiro — mine shaft
titanio — titanium
titanita — sphene, titanite
toba — calcareous tufa, travertine, volcanic tuff
tonelada — ton
topacio — topaz
torbernita — torbernite
torio — thorium
translúcida — translucent
transparente — transparent
tranvía — tramway
traquito — trachyte
travertino — travertine
tremolita — tremolite
Triásico — Triassic
tridimita — tridymite
triturador — crusher
troilita — troilite
trona — trona
tubo — tube, pipe
tulita — thulite
tungsteno — tungsten
turba — peat
turgita — turgite
turmalina — tourmaline
turquesa — turquois

U - V

uralita — uralite
uraninita — uraninite
uranio — uranium
urano — uranium
vacío, - ía — empty
válvula — valve
valle — valley
vanadinita — vanadinite
vanadio — vanadium
variscita — variscite
varisita — variscite
vapor — steam
vela — candle
velador — watchman
velocidad — velocity
ventilación — ventilation
vermiculita — vermiculite
vesuviana — idocrase, vesuvianite
vesuvianita — vesuvianite
veta — vein, fissure vein
vidrio — glass
vítreo — vitreous
volcán — volcano
volframita — wolframite
vulcanio — igneous

W - X - Y - Z

wad — wad
willemita — willemite
witerita — witherite
wolframita — wolframite
wollastonita — wollastonite
wulfenita — wulfenite
wurtzita — wurtzite
yacimiento — ore deposit
yeso — gypsum
yeso var. alabastro — gypsum var. alabaster
yeso var. selenita — gypsum var. selenite
yunque — anvil
zafir — sapphire
zafiro — sapphire
zinc — zinc
zirconio — zirconium
zoicita — zoisite
zona — zone, district

Selected List of Minerals of Mexico

(Including Baja California)

The following list of minerals, including gem minerals, has been compiled mainly from the English literature. Had Spanish literature been consulted, it undoubtedly would have been longer. This list contains about 376 mineral species. A question mark (?) after a name indicates a doubtful species or unconfirmed occurrence.

acanthite
achrematite (?)
actinolite
actinolite pseudomorphous
  after pyroxene
adamite
adamite, var. cuproadamite
adularia (orthoclase)
agate (also see quartz)
aguilarite
aikinite
alabandite
alabaster
alamosite
albite
albite, var. cleavelandite
allemontite
allophane
almandite
alunite
alunogen
amesite
amethyst (quartz)
amphibole group
  actinolite
  cummingtonite
  hornblende
  tremolite
analcime (analcite)
anatase
andalusite
andalusite,
  var. chiastolite
andesine
andradite
anglesite
anhydrite
ankerite
annabergite
antimony, native
antimony ocher
antlerite
apatite
apatite, var. collophane
apophyllite
aquamarine (beryl)
argentite
argentite, cuprian
  (jalpaite)
arsenic, native
arseniosiderite
arseniosiderite pseudomorphous after scorodite
arsenopyrite
arsenopyrite, auriferous
atacamite
augite
augite, var. leucaugite (?)
aurichalcite
austinite
axinite
azurite
barcenite (discredited)
barite
barthite (same as austinite)
beidellite (discredited)
berthierite
bertrandite
beryl
  var. aquamarine
  var. morganite
bindheimite
biotite
birnessite
bismite
bismuth, native
bismuth ocher (oxides)
bismuthinite

bismutite
boleite
bornite
boulangerite
bournonite
braunite
brewsterite (?)
brochantite
brookite
bromyrite
bronzite (see enstatite)
brucite
burkeite
bustamite
bystromite
bytownite
calcite
calcite, var.
manganocalcite
calomel
carbonate-apatite
(dahllite)
carmenite (discredited)
carminite
carnelian (quartz)
carnotite
carphosiderite
cassiterite
celadonite
celestite
cerargyrite
cerussite
cerussite, argentiferous
chabazite
chabazite, var. phacolite
chalcanthite
chalcedony (quartz)
chalcocite
chalcophanite
chalcopyrite
chalcopyrite, argentiferous
chalcopyrite, auriferous
chalcostibite
chiastolite (see andalusite)
chlorite
chromite
chrysocolla
chrysolite (see olivine)
chrysotile
cinnabar
cleavelandite (see albite)
clinochlore
clinozoisite
cobaltite
cobaltocalcite
coccinite
cocinerite (?)
collophane (see apatite)
columbite
conichalcite
copiapite
copper, native
coquimbite
cordierite (iolite) (?)
coronadite
corundum
cosalite
cosalite, argentian
covellite
crednerite
cristobalite
cryptomelane
cumengite
cummingtonite
cuprite
cuproadamite (see adamite)
cuprodescloizite (see
descloizite)
cuprotungstite
dahllite (see carbonate-
apatite)
danburite
datolite
davidite
daubreelite
delafossite
denningite
descloizite
descloizite, var. cupro-
descloizite
(ramirite or
tritochorite)
diaphorite
dickite

digenite
diopside
diopside, var. salite
dioptase
dolomite
domeykite
dravite (see tourmaline)
duftite
dumortierite
durangite
dussertite
eglestonite
embolite
emmonsite
enargite
endlichite (see vanadinite)
enstatite
enstatite, var. bronzite
epidote
epsomite
erythrite
essonite (see grossular)
famatinite (?)
fayalite
feldspar group
   microcline
   orthoclase
   plagioclase
fergusonite
fluorapatite
fluorite
fowlerite (?)
freibergite (see tetrahedrite)
freieslebenite
galena
galena, argentiferous
garnet group
   almandite
   andradite
   grossular
   pyrope
   spessartine
garnierite (?)
gehlenite
germanite (?)
gersdorffite
glauberite
glaucodot
goethite
goethite ("limonite") pseudo morphous after pyrite
gold, native
goslarite
graphite
greenockite
grossular
grossular, var. essonite (hessonite)
grossular, var landerite (rosolite, xalostocite)
guadalcazarite, see metacinnabar
guanajuatite
gypsum
gypsum, var. alabaster
gypsum, var. selenite
halite
halloysite
halotrichite
hanksite (?)
harmotome
hedenbergite
hedenbergite, var. manganhedenbergite
hematite
hematite, var. martite pseudomorphous after magnetite
hematite, var. specularite
hematite, var. turgite
hemimorphite
herderite
hessite
hessonite (see grossular)
heulandite
hidalgoite
hidalgoite, ferrian
hillebrandite
hisingerite
hollandite
hornblende, "basaltic"
hornblende, common
huebnerite

hyalite, see opal
hydrobiotite
hydrozincite
hypersthene
iddingsite
idocrase (vesuvianite)
illite
ilmenite
ilsemannite
ilvaite
indicolite, (see tourmaline)
inesite
iodyrite
iolite (see cordierite)
jadeite. Found as artifacts only, none found in situ.
jalpaite (see argentite)
jamesonite
jarosite
jasper (see quartz)
johannsenite
kaolin group
- allophane
- dickite
- halloysite
- kaolinite

kaolinite
kermesite
klaprothite
knebelite
kobellite
krausite
labradorite
landerite (see grossular)
langite
laumontite
lead, native
legrandite
lepidolite
leucaugite
leucoxene
"limonite"
livingstonite
loellingite
lucianite
mackayite
magnesite
magnetite
malachite
manganhedenbergite, (see hedenbergite)
manganite
manganocalcite (see calcite)
manganosiderite
marcasite
marmatite (see sphalerite)
martite (see hematite)
massicot
matildite
maucherite
mazapilite (probably identical with arseniosiderite)
melaconite
melanterite
meneghinite
mercury, native
merwinite
metacinnabar
metacinnabar
- var. guadalcazarite
- var. onofrite

metavoltine
miargyrite
mica group - see
- lepidolite
- muscovite
- phlogopite
- vermiculite

microcline
millerite
mimetite
minium
mirabilite
mixite
molybdenite
molybdite
monazite
monticellite
montmorillonite
montroydite
morganite (see beryl)
mosesite

mottramite
mountain leather
(amphibole asbestos)
murdochite
muscovite
muscovite, var. sericite
nagyagite
natrojarosite
natrolite
natron.
naumannite
neodigenite
nepheline
niccolite
nickel, native
niter
nontronite
nsutite
oligoclase
olivenite
olivine (chrysolite)
olivine, var. peridot
onofrite (see metacinnabar)
opal, common
opal, precious
opal, var. hyalite
ordonezite
orpiment
orthoclase
orthoclase, var. adularia
orthoclase, var. sanidine
paradamite
paraguanajuatite
paratellurite
pearceite
penninite
pentlandite
percylite
peridot (see olivine)
petzite
phacolite (see chabazite)
pharmacolite
phlogopite
phosgenite
pitchblende (see uraninite)
plagioclase - see
albite
andesine
bytownite
labradorite
oligoclase
plattnerite
plumbojarosite
polyargyrite (?)
polybasite
portlandite
powellite
prehnite
priceite
prochlorite
proustite
pseudoboleite
pseudobrookite
psilomelane
pyrargyrite
pyrite
pyrite, auriferous
pyrite, cupriferous
pyrochroite
pyrolusite
pyromorphite
pyrope
pyrophyllite
pyroxene group - see
augite
diopside
enstatite
hedenbergite
hypersthene
jadeite
pyrrhotite
pyrrhotite, var. triolite
quartz
quartz pseudomorphous after
aragonite
barite
calcite
quartz with inclusions of
hematite
tourmaline
(tourmalinated)

quartz, var.
- rose
- smoky
- chrysocolla stained (silica)
- agate
- amethyst
- carnelian
- chalcedony
- jasper
- tigereye(?)

ramirite (see descloizite)
rammelsbergite
ramsdellite
raumonite (?)
realgar
rhodite
rhodochrosite
rhodonite
rosasite
rose quartz (see quartz)
rosolite (see grossular)
rubellite (see tourmaline)
rutile
safflorite
salite (see diopside)
sanidine (see orthoclase)
saponite
sartorite
sauconite (?)
scapolite
scheelite
schorl (see tourmaline)
scorodite
selenite (see gypsum)
selenium, native
sepiolite
sericite (see muscovite)
serpentine
siderite
sillenite
sillimanite
silver, native
smaltite
smithsonite
smithsonite, cobaltiferous
smoky quartz (see quartz)
specularite (see hematite)
spessartite
sphalerite
sphalerite, var. marmatite
sphene (titanite)
spinel
spiroffite
spodumene
spurrite
stephanite
sternbergite
stibiconite
stibiobismuthinite
stibnite
stilbite
stromeyerite
strontianite
sulfur, native
svabite
talc
"tapalpite" a mixture of tetradymite and argentite
teepleite
tellurite
tellurium, native
tellur-jarosite
tennantite
tenorite
terlinguaite
tetradymite
tetrahedrite
tetrahedrite, argentian (freibergite)
tetrahedrite, auriferous
thermonatrite
thmonsonite

thomsonite
thulite (see zoisite)
tiemannite
titanite (see sphene)
topaz
torbernite
tourmalinated quartz (see quartz)

tourmaline
tourmaline, var.
green
bicolor
dravite
indicolite
rubellite
schorl
watermelon
tremolite
tridymite
triolite (see pyrrhotite)
tritochorite (see descloizite)
trona
tungstite
turgite (see hematite)
turquois
tuxtlite
uralite (see actinolite)
uraninite (pitchblende)
uranophane
valencianite (see orthoclas orthoclase)
valentinite
vanadinite
vanadinite, var endlichite)
variscite
vermiculite
vesuvianite (see idocrase)
violarite
voltaite
wad
wagnerite
whitlockite
willemite
wolframite
wollastonite
wulfenite
wurtzite
xalostocite (see grossular)
xenotime
xonotlite (xonaltite)
zeolite group - see
analcime
chabazite
heulandite
laumontite
natrolite
stilbite
zinkenite
zircon
zoisite
zoisite, var. thulite

Minerals Named For Mexican Mines or Localities

alamosite
cocinerite
cosalite
cristobalite
durangite
guadalcazarite
guanajuatite
hidalgoite
jalpaite
lucianite
onofrite
paraguanajuatite
valencianite
xalostocite
xonotlite

Mexican States and Abbreviations

Aguascalientes — Ags.
Baja California, Norte — B.C.N.
Baja California, Sur (territory) — B.C.S.
Campeche — Camp.
Chiapas — Chis.
Chihuahua — Chih.
Coahuila — Coah.
Colima — Col.
Distrito Federal — D.F.
Durango — Dgo.
Guanajuato — Gto.
Guerrero — Gro.
Hidalgo — Hgo.
Jalisco — Jal.
México — Méx.
Michoacán — Mich.
Morelos — Mor.
Nayarit — Nay.
Nuevo León — N.L.
Oaxaca — Oax.
Puebla — Pue.
Querétaro — Qro.
Quintana Roo — Q.R.
San Luis Potosí — S.L.P.
Sinaloa — Sin.
Sonora — Son.
Tabasco — Tab.
Tamaulipas — Tamps.
Tlaxcala — Tlax.
Veracruz — Ver.
Yucatán — Yuc.
Zacatecas — Zac.

Mineral Species First Discovered in the Republic of Mexico

Numbers in parentheses denote the section, as shown by the Index Map, in which the locality will be found.

ACHREMATITE - from Guanacere, Chihuahua. Ill defined. Probably an invalid species. (2)

AGUILARITE - from San Carlos mine, Guanajuato, Guanajuato. (2)

ALAMOSITE - from Alamos, Sonora. (1)

BARCENITE - from Huitzuco, Guerrero. Needs verification to become a valid species. (5)

BYSTROMITE - from La Fortuna mine, El Antimonio district, Sonora. (1)

COCCINITE - from Casas Viejas, near San Jose Iturbide, Guanajuato. (2)

COCINERITE - from Cocinera mine, Ramos, San Luis Potosi. (3)

COSALITE - from Cosala, Sinaloa. (1)

CRISTOBALITE - from Cerro San Cristobal, near Pachuca, Hidalgo. (4)

DURANGITE - from Barranca mine (tin), 18 miles northeast of Coneto, Durango. (2)

GUADALCAZARITE (a variety of metacinnabar) - from Guadalcazar, San Luis Potosi. (3) A zincian metacinnabar.

GUANAJUATITE - from Santa Catarina and La Industrial mines, Sierra Santa Rosa, near Guanajuato, Guanajuato. (2)

HIDALGOITE - from San Pascual mine, Zimapan district, Hidalgo. (3)

HILLEBRANDITE - from Terneras mine, Velardena district, Durango. (2)

IODYRITE - from Albarradon, near Mazapil, Zacatecas. (2)

JALPAITE (a variety of argentite) - from Jalpa, Zacatecas. (2) A cuprian argentite.

KRAUSITE from Santa Maria mine, Velardena district, Durango. (2)

LEGRANDITE - from Flor de Pena mine, Lampazos, Nuevo Leon. (3)

LIVINGSTONITE - from Huitzuco, Guerrero. (5)

LUCIANITE - from Hacienda Santa Lucia, near Mexico City. (4)

ONOFRITE (a variety of metacinnabar) - from San Onofre, near Plateros, northwest of Zacatecas, Zacatecas. (2) A selenian metacinnabar.

ORDONEZITE - from Santin mine, Cerro de las Fajas, Santa Catarina; Guanajuato. (2)

PARADAMITE - from Ojuela mine, near Mapimi, Durango. (2)

PARAGUANAJUATITE - from Santa Catarina mine, Sierra de Santa Rosa, near Guanajuato, Guanajuato. (2)

PARATELLURITE - from Cananea, Sonora. (1)

PERCYLITE - locality given only as State of Sonora. (1)

SILLENITE - from Durango, Durango. (2)

SPIROFFITE - from Moctezuma, Sonora. (1)

SPURRITE - from Velardena district, Durango. (2)

TRIDYMITE - from Cerro San Cristobal, near Pachuca, Hidalgo. (4)